Education Outrage

Education Outrage

Roger C. Schank

Constructing Modern Knowledge Press

These articles originally appeared as blog posts on:
Education Outrage (educationoutrage.blogspot.com) and The Pulse: Education's Place for Debate (no longer online).

Constructing Modern Knowledge Press

Torrance, CA, USA

www.CMKPress.com

EDU034000 EDUCATION / Educational Policy & Reform

EDU042000 EDUCATION / Essays

ISBN: 978-0-9891511-3-9

Editor: Sylvia Libow Martinez

Copy editors: Alice Richardson, Vivian Martinez

Cover design: James Todd

Table of Contents

Foreword ix

Chapter 1: Why?

Why do we still have schools? 2
Why is school screwed up? 6
The 7th P 9
Why students cheat 11
Why do we teach what we teach? 13
Why students major in history and not science 14
Why don't we encourage schools to adapt to kids rather than the other way around? 16
Why do we give lectures? Why does anyone attend them? 19
Why our education system is the way it is: The Tillman Story 22
Why not use the principles of prison reform to help schools? 24
Why educate the elite? A lesson from Yale and George Bush. 25
The top ten mistakes in education: twenty years later 27

Chapter 2: Teaching the Wrong Things

Wrong problem, wrong solution 30
In defense of what doesn't work 32
The curriculum deciders 34
Benito Juarez, Don Quixote, and Mexico City street kids 35
Stop pushing math in the name of reasoning 38
Why do we let the general public decide what should be taught in school? 41
Why textbooks suck 43
Sacred truths 45
An honest back to school message to students 47
NO to subjects and NO to requirements 49
The politics of history 51
The peril of passion 52
National standards are absurd 54
Does anyone care that kids all over the world think school is useless? 56

Chapter 3: Teaching the Wrong Way

Cavemen didn't have classrooms 60
The library metaphor 64
How to build a culture of illiteracy 66
The history teacher or the football coach? 67

Can you tell if it is school or prison? 70
Why can't we express emotion properly? A call for new school standards. 71
Teachers' despair: we cannot afford to be focused on training intellectuals 75
The myth of information retention 78
Exposure, cultural literacy and other myths of modern schooling 81
Reading and Math are very important, but I am not sure why: a message for those who can't see very well 84
Reading is no way to learn 88

Chapter 4: Technology Saves the World!

Games to the rescue! 92
Alex Trebek: hero of vocabulary preparation 94
The online learning disaster 95
Stanford decides to be Wal-Mart 96
Greed is not disruption 98
Princeton Professor teaches Coursera course; you must be kidding me! 100
Efficiency is not reform 102
MOOCs, the XPRIZE, and other things that will never change education 104
MOOCs: The New York Times gets it wrong again; Europe is not lagging behind the U.S. 107
Why are universities so afraid of online education? 110
More online nonsense: Starbucks and Arizona State agree to do nothing useful 113
Students: Be very afraid of online degree programs, especially if Pearson had anything to do with them 115
What online education should be 118

Chapter 5: Everyone Should Go To College?

Everyone must go to college. Does anyone ever ask why? 122
Should I go to a "hot" college? 124
The Big Lie: teaching never matters in university rankings 126
Stop cheating undergraduates of a useful education 128
How academic research has ruined our education system 130
Humanities are overrated 134
"What should I go to school for?" 136
Engaged learning 138
Confused about what college is about? So are colleges. 139
Preparing for a fictitious college 142
Only Harvard and Yale lawyers on the Supreme Court? 143
Please don't make me be a dentist! 144
Students: Life isn't a multiple choice test. Have some fun. 146
How do you know if you are college-ready? 149
Parents, relax. Your kid will get into college; the question is whether or not to go. 150

Chapter 6: Death to the Standardistas!

Death to the Standardistas! 154
What is wrong with trying to raise test scores? 156
Chinese do better on tests than Americans! Oh my God, what will we do? 158
The World Cup of testing 161
Frank Bruni thinks kids are too coddled. I think kids are too tested. 165
Measurement in preschool? Measure this! 167
"Life is a series of tests." What a load of nonsense. 169
American business hasn't a clue 170
Measure or die 172
Now I am disgusted! 174

Chapter 7: Dear Mr. Obama (and Other Politicians)

Mr. Obama wants big ideas? Here are 10 in education. 176
Duncan speaks; kids lose 177
A message to Bachmann, Duncan, and every other politician who thinks he or she knows how to fix education 179
Just the facts, ma'am 181
Thank you Arne, Bill, and Pearson for making this teacher so miserable 183
Why are you proud, Mr. Mayor? 185
More college graduates? Say it ain't so Mr. President. 187
Yes, Mr. Obama, money is the answer 188
What were your test scores, Mr. Obama? 189
The school of random facts 190
Hooray for the Democrats! Hooray for more accountability! 191
Why politicians and rich guys won't reform education 194
Public schools: where poor kids go to take tests 196
Duncan and Obama are actively preventing meaningful education change 197
Free community college? How about we fix high school, Mr. Obama? 198

Chapter 8: The New York Times (and Others) – Wrong Again

The New York Times and Nick Kristof want mass education. I want individualized education. 202
Spinning test prep into "choice" 204
The New York Times on the GED – wrong again 206
When the New York Times obsesses about math, every kid loses 208
Tom Friedman; wrong again, this time about education 210
Don't worry about Artificial Intelligence, Stephen Hawking 211
The misreporting of science by The New York Times and others 214
I translate Bill Gates dumb remarks on education 216
Bill Gates: wrong again about rating teachers 219
Madrassas, indoctrination, education, and Kristof 220

More absurdity from the New York Times and Nick Kristof 223
The fake choice: preschool or prison 225
Measuring teachers means education reform? You have got to be kidding! 227
Fared Zakaria and Ivy League graduates keep defending the liberal arts, but clearly the liberal arts didn't teach them to think 229

Chapter 9: Milo

My grandson visits 234
A drum should be a drum 236
Milo goes to kindergarten. We need to fix this fast. 240
Milo and the Park Slope parents meet Common Core; everybody loses (except Bill Gates and Big Pharma) 241
Milo's mom fights the Common Core 244
Milo takes a computer class and eventually will learn PowerPoint. Yippee! 247
Why can't school be more like camp? 249

Chapter 10: How To Fix Education

It's time for a change 252
A conversation about learning 255
We learn by talking 260
Persuasive arguments 263
Change just one thing 265
The old university system is dead – time for a professional university 266
What cognitive science tells us about how to organize school 268
Start by asking students what they want to learn 270
Pro-Choice: here's how to fix high schools 272
High school: Out with the old and in with the new 275
Cash in those chips 277
It's an old story – learning hasn't changed 278
How to fix the "STEM crisis" 281
To fix a country, fix education 283
A summer to-do list for students who hate (or love) high school 285

Original Publication Dates 289

About the Author 295

Foreword

If teachers or parents won't stand between kids and the madness who will?

One of my all-time favorite books is 1988's *Holidays In Hell*. In that book, gonzo journalist and satirist P.J. O'Rourke invites readers to tag along as he vacations in some of the worst places in the world. *Education Outrage* shares much with that earlier volume. Instead of war-torn Beirut, the systemic incompetence of East Germany, or the depravity of Heritage USA, Roger Schank leads readers on a tour of your child's homework backpack, irrelevant curricula from kindergarten to college, destructive education policy, and the violent wasteland that is the *New York Times* opinion page.

Neither Schank nor O'Rourke care whether you agree with them. They upend conventional wisdom with a keen analytical eye and razor-sharp wit. It does not matter if I share these authors' politics or worldview, they make me see things I may have missed and recognize that things need not be as they seem. While reading this book, shock may quickly turn to awe, accompanied by belly laughs. I hope that a fresh perspective will cause you take action on behalf of your own child or those you teach.

As you confront "outrageous" ideas between these covers do not succumb to the notion that Professor Schank is playing "devil's advocate." He is the *learner's* advocate in an age when the rights, needs, and desires of students are dismissed, suppressed, or punished in support of conventional wisdom propagated by folks lacking discernible wisdom. Roger Schank speaks truth to power in an asymmetric war being waged on young people, learning, and common sense. He never punches down. His animus is aimed at those in the corridors of power whose uninformed opinions ignore modernity, shame teachers, enforce the indefensible, and rob kids of a joyous childhood.

Roger Schank is no mere crank or dilettante opining about how school should be for other people's children. He is an expert learner and an expert on learning. Schank is a distinguished university professor, mathematician, linguist, computer scientist, artificial intelligence pioneer, entrepreneur, CNBC television host, software developer, author, parent, grandparent, and "old guy" softball player. Those who seek to educate young people should embrace his remarkable experience, wisdom, and expertise. Sadly, unqualified is the new qualified when it comes to setting education policy in the United States and across the globe.

Schank would be the first to tell you that he has better things to do with his time than fix schools.[1] However, this book should inspire those working in schools to reflect upon their personal practice.

1 He's not really a fan of schools in the first place.

This book organizes and edits several years worth of blog posts by Roger Schank in order to preserve them for posterity. A current event mentioned in an essay may have faded from memory, but those of us who work every day to make the world a better place for learners know that the conditions Roger describes are getting more hellish, not less.

Those of us who know better must do better. Reading this book will help you know more. May you find the courage to do the right thing for all of our young people.

Gary S. Stager, Ph.D.
Educator and Publisher: Constructing Modern Knowledge Press

Chapter 1: Why?

Why do we still have schools?

People get used to the institutions that have been a part of their lives. This is especially true of institutions that have been around for many generations, and of institutions whose purpose is seen as doing something worthwhile. Add into the mix that the absence of that institution in certain places around the world is always correlated with poverty, and you have a situation where no one ever questions the value of that institution.

Nevertheless, I will ask a heretical question: Why do we have schools? Instead of answering this question by listing all the good things that schools provide, which anyone can do, I will turn the question around: What is bad about having schools?

Here is my list of answers:

Competition
Why should school be a competitive event? Why do we ask how a kid is doing in school? Learning in life outside of school is not a competitive event. We learn what we choose to learn in real life.

Stress
When six-year-olds are stressed about going to school you know that something is wrong. Is learning in real life stressful? Stress can't be helping kids learn. What kid wouldn't happily skip school on any given day? What does this tell us about the experience?

Right Answers
School teaches that there are right answers. The teacher knows them. The test makers know them. Now you have to know them. But, in real life, there are very few right answers. Life isn't mathematics. Thinking about how to behave in a situation, planning your day or your life, plotting a strategy for your company or your country – no right answers.

Bullying and Peer Pressure
You wouldn't need "say no to drugs and cigarettes" campaigns if kids didn't go to school. In school there are always other kids telling you how to dress, how to act, how to be cool. Why do we want kids' peer groups to be the true teachers of children? Being left out terrorizes children. Why do we allow this to happen by creating places that foster this behavior?

Stifling of Curiosity

Isn't it obvious that learning is really about curiosity? Adults learn about things they want to learn about. Before the age of six, prior to school, one kid becomes a dinosaur specialist while another knows all about dog breeds. Outside of school, people drive their own learning. Schools eliminate this natural behavior.

Subjects Chosen For You

Why algebra, physics, economics, and U.S. history? Because those subjects were pretty exciting to the President of Harvard in 1892. And, if you are interested in something else – psychology, business, medicine, computers, design? Too bad. Those subjects weren't taught at Harvard in 1892. Is that nuts or what?

Classrooms

If you wanted to learn something and had the money, wouldn't you hire someone to be your mentor, and have them be there for you while you tried out learning the new thing? Isn't that what small children have, a parent ready to teach as needed? Classrooms make no sense as a venue for learning, unless of course you want to save money and have 30 (or worse, hundreds of) students handled by one teacher. Once you have ratios like that, you have to teach by talking and then hope someone was listening, so then you have to have tests. Schools cannot work as places of learning if they employ classrooms. And, of course, they pretty much all do.

Grades

Any professor can tell you that students are pretty much only concerned with whether what you are telling them will be on the test and what they might do for extra credit. In other words, they want a good grade. If you tell them that 2+2=5 and it will be on the test, they will tell you that 2+2=5 if it means getting a good grade. Parents do not give grades to children and employers do not give grades to employees. They judge their work and progress, for sure, but not by assigning numbers to a report card.

Certification

We all know why people attend college. They do so primarily to say they are college graduates so they can get a job or go on to a professional school. Most don't care all that much about what hoops they have to go through. They do what they are told. Similarly, students try to get through high school so they can go on to college. As long as students are not in school to get an education, you can be pretty sure they won't get one. Most of our graduates have learned to jump through hoops, nothing more.

Confined Children

Children like to run around. Is this news to anyone? They have a difficult time sitting still and they learn by trying things out and asking questions. Of course in school, sitting still is the norm. So we have come up with this wonderful idea of ADHD, i.e. drug those who won't sit still into submission. Is the system sick or what?

Academics Viewed as Winners
Who are the smartest kids in school? The ones who are good at math and science, of course. Why do we think that? Who knows? We just do. Those who are good at these subjects go on to be professors. So those are certainly the smartest people we have in our society. Perhaps they are. But, I can tell you from personal experience that our society doesn't respect professors all that much, so something is wrong here.

Practical Skills Not Valued
When I was young there were academic high schools and trade high schools. Trade high schools were for dumb kids. Academic high schools were for smart kids. We all thought this made sense. Except that there are a lot of unemployed English majors and a lot of employed airplane mechanics. Where did we get the idea that education was about scholarship? This is not what Ben Franklin thought when our system was being designed, but he was outvoted.

The Need to Please Teachers
People who succeed at school are invariably people who are good at figuring out what the teacher wants and giving it to them. In real life there is no teacher to please and these "grade grubbers" often find themselves lost. When I did graduate admissions, if a student presented an undergraduate record with all A's I immediately rejected him. There was no way he was equally good at, or equally interested in, everything. (Except pleasing the teacher.) As a professor, I had no patience for students who thought that telling me what I just told them was the essence of academic achievement.

Self-Worth Questioned
School is full of winners and losers. I graduated number 322 in my high school class (out of 678). Notice that I remember this. Do you think this was good for my self-esteem? Even the guy who graduated number 2 felt like a loser. In school, most everyone sees themselves as a loser. Why do we allow this to happen?

Politicians in Charge
Politicians demand reform, but they wouldn't know reform if it hit them over the head. What they mean is that school should be like they remember it, rather than how it is now, and they will work hard to get you to vote for them, to give them money to restore the system to the awful state it was always in. Politicians, no matter what party, actually have no interest in education at all. An educated electorate makes campaigning much harder.

Government Use of Education for Repression
As long as there have been governments, there have been governments who want people to think that the government (and the country) is very good. We all recognize this tendency in dictatorships that promote the marvels of the dictator and rewrite history whenever it is convenient. When you point out that our government does the same

thing you are roundly booed. We all know that the Indians were savages, that Abraham Lincoln was a great President and that we are the freest country on earth. School is about teaching "truth."

Discovery Not Valued

The most important things we learn, we teach ourselves. This is why kids have trouble learning from their parents' experience. They need their own experiences to ponder and to learn from. We need to try things out and see how they go. This kind of learning is not valued in school because it might lead to, heaven forbid, failure, and failure is a really bad word in school. Except failure is how we learn, which is pretty much why school doesn't work.

Boredom Ignored

"Boredom is a bad thing." We drug bored kids with Ritalin so they will stop being bored. All of my best work has come when I was most bored and let my mind wander. It is odd that we keep trying to prevent this from happening with kids. Lots of TV, that's the ticket.

Major Learning by Doing Mechanism Ignored

And last but not least, scholars from Plato to Dewey have pointed out that people learn by doing. That is how we learn. Doing. Got it? Apparently not. There is very little doing in schools. Unless you count filling in circles with number-two pencils as doing.

Why is school screwed up?

It is very difficult to think about replacing sacred institutions. The only way I know how to think about it is as a thought experiment. Imagine that we live in a different world, maybe a colony in the first century, and ask yourself how we might educate our children in this environment, pretending that for some reason, schools are the one thing we cannot build. As we think about this, we must not assume that what we teach in schools now needs to be taught in some other way. We simply need to ask what one should one teach children, without assuming that what we have been teaching today is the right thing.

To put this another way, the right question to ask is – what do we need to be able to do in order to function in the world we inhabit? The next question is, of course, how would we teach children to do those things?

Now admittedly I am prejudicing the answer here by simply leaving out the word "know." The usual question is what should children "know"? It is this question that leads crazies to make lists of things every third grader should know, and allows school boards to create lists of facts that students need to be tested on. So, let's leave that word out of the discussion and see where it gets us.

A good place to start is to ask what a highly functioning adult can do and, moreover, has to be able to do in order to live in this world. While we ask this question the phrase "twenty-first-century skills" will not come up. Every time that phrase comes up the answer somehow turns out to include algebra and calculus and science, which, the last I heard, were nineteenth-century skills too.

In fact let's not talk about particular centuries at all. To see why, I want to diverge for a moment into a discussion of the maritime industry, a subject with which I have become more fascinated over the years. What did a mariner from Ancient Greece have in common with his modern counterpart in terms of abilities?

The answer is obsession with weather, ship maintenance, leadership and organization, navigation, planning, goal prioritization, and handling of emergencies.

What effective mariners from ancient times would have in common with those of today is an understanding of how to operate their ships, the basic laws of weather, tides, navigation and other relevant issues in the physical world, and an ability to make good decisions when circumstances are difficult. They would also have to know how to get along with fellow workers, how to manage those who report to them, as well as basic laws of commerce and defense.

In fact, the worlds they inhabit, from an educational point of view – that is, from thinking about what to teach and how to teach it – would be nearly identical except for one thing: how to operate and maintain the equipment. Their ships were, of course, quite different.

So, let's re-formulate this question that seems to haunt every modern day pundit on education (usually politicians or newspaper peoples). "What are twenty-first-century skills?" can be transformed (for mariners) into "What does a twenty-first-century mariner need to be educated about that his Ancient Greek counterpart was not educated about?"

The answer, it seems obvious to me, is twenty-first-century equipment and procedures: engines, navigation devices, particular political situations, computers and so on. But – and this is an important but – none of this stuff is the real issue in the education of a mariner. The real issue is decision-making. What one has to make a decision about is secondary to the issue of knowing how to make a decision at all.

You can learn about a piece of equipment or a procedure by apprenticeship. Start as a helper and move on gradually to being an expert. But this is not what school emphasizes. School typically attempts to intellectualize these subjects. Experts write books about the theory of how something works and the next we know, schools are teaching that theory as a prelude to actually doing the work. Scholarship has been equated with education. You do not have to know calculus to repair an engine. You might want to know calculus to design an engine, but that is no excuse for forcing every engineer to learn it. Similarly, you do not have to know theoretical physics to master the seas. Mariners do know physics of course – practical physics about load balancing, for example – but they do not have to know how to derive the equations that describe it.

What I am saying here about the shipping Industry holds true for every other area of life. Twenty-first-century skills are no different than first-century skills. Interestingly, Petronius, a first-century Roman author, complained that Roman schools were teaching "young men to grow up to be idiots, because they neither see nor hear one single thing connected with the usual circumstances of everyday life." In other words, schools have always been about educating the elite in things that don't matter much to anyone. This is fine as long as the elite don't have to work.

But, today the elites have extrapolated from what they learned at Harvard and decided that every single school child needs to know the same stuff. So, they whine and complain about math scores going down without once asking why this could possibly matter. Math is not a twenty-first-century skill any more than it was a first-century skill. Algebra is nice for those who need it and useless for those who don't. Skill in mathematics is certainly not going to make any industrial nation more competitive with any other no matter how many times our "experts" assert that it will. One wonders how politicians can even say this junk, but they all do.

Why?

My own guess is that, apart from the fact that they all took these subjects in school (and were probably bad at them – you don't become a politician or a newspaper person because you were great at calculus), there is another issue. They don't know what else to suggest.

Thinking about the first century will help us figure out what the real issues are. People then and people now have had to learn how to function in the world they inhabit. This means being able to communicate, get along with others, function economically and physically, and reason about issues that confront them. It didn't mean then, and doesn't mean now, science and mathematics, at least not for 95% of the population.

So, what should we teach students?

Offer choices. Stop making lists of what one must know and start putting students into situations where they can learn from experience while attempting to accomplish goals that they set out for themselves, just as people did before there were schools. Education has always been the same: learning from experience with the help from wiser mentors. School has screwed that all up and it is time to go back to basics.

The 7th P

I used to say that everything evil in education starts with the letter P. There were six Ps. But now I have found a 7th.

To review, the 6 Ps are:

1. **Parents** – who oppose all change and want school to be like they imagine it was in their day.
2. **Publishers** – who spend all that money on wrong-headed textbooks and do their best to keep new ideas away.
3. **Press** – who print minute test score differences as if they are world-shaking events, causing everyone to panic.
4. **Politicians** – who really don't give a hoot about education and just like to say how accountable everyone is because of their silly tests and standards.

And Princeton (twice)

5. **Princeton** – as in any top university that decides on which courses and which tests all students must pass, thus making it very difficult to innovate in high school.
6. **Princeton** – as in the Educational Testing Service and all the other testing companies getting rich on killing our schools.

But alas, I have realized that there is a 7th P. My realization reminded me of Marshall McLuhan's remark, "I don't know who discovered water but I'm pretty sure it wasn't a fish." And also of the cartoon Pogo's observation that "We have met the enemy and he is us."

The 7th P? Professors

I knew I quit the university for a reason. The other day I was reminded why. I was running one of my design meetings for my new learn-by-doing virtual high. We were working on a full year curriculum in engineering and with the help of Boeing we were focusing on aerospace. Normally I decide who goes to these design meetings but in this case Boeing brought its own engineers, which was fine, and a professor they knew. The professor was concerned that students arrived at his university not knowing enough of the basics, and he was hoping that this meeting was about teaching math and science better so that his job would be easier when the students arrived on campus.

Of course, this was the exact opposite of what I was trying to do. The reason school is so bad is that everyone is learning stuff they hate just so some small percentage of students will be ready to take a university course of some sort. Everyone is bored to death so this guy's life will be easier and he won't have to teach the basics.

This idea is so ubiquitous that I never noticed it until I was far enough away. Universities dictate curricula to high schools to make professor's lives easier. If everyone takes physics and calculus and most never use it, well, professors claim it was good for the students anyway when in fact it was only good for making sure professors didn't have to teach it in college. As long as professors don't have to teach the basics, it is okay that high school students are forced to study stuff they will never use. We have ruined an entire generation of high school students who don't like learning and think the subject matter is irrelevant because professors only want to teach the good stuff.

We sacrifice the joy of learning for an entire generation so professors can have an easier time teaching incoming students.

Why students cheat

Lately there has been a great deal written about student cheating. Today there was an editorial in the *New York Times*, which always gets education wrong.[1] The question of why students cheat is usually answered by mentioning that it is the fault of the Internet, or by listing the big three reasons, which are:

- The pressure to get good grades
- That students are lazy and didn't do the required work
- That students thought they could get away with it

The *New York Times* editorial quotes a professor who says "This represents a shift away from the view of education as the process of intellectual engagement through which we learn to think critically and toward the view of education as mere training. In training, you are trying to find the right answer at any cost, not trying to improve your mind."

The editorial goes on to mention that, "More than half the colleges in the country have retained services that check student papers for material lifted from the Internet and elsewhere." And then the writer adds, "Parents, teachers and policy makers need to understand that this is not just a matter of personal style or generational expression. It's a question of whether we can preserve the methods through which education at its best teaches people to think critically and originally."

I wonder if there could be a better explanation of why students cheat? Perhaps the answer is that the professors and their universities encourage students to cheat. Let me explain.

Consider the Motor Vehicle bureau's approach to education. Why aren't we hearing about rampant student cheating in driver's license exams? Perhaps there is cheating on the written tests, I don't know. But I am pretty sure there isn't cheating on the actual driving test. Why not? Because that test is a test of performance ability, not competence. The driver's test tests to see if you can actually do something and there is a person looking to see if you can do it.

Now let's think about the university model of education. Universities don't actually ask professors to see if their students can do anything in a one-on-one encounter as the motor vehicle bureau does. Why not?

Because the universities are cheating. They are cheating in two ways. First, they are claiming that education consists of one professor talking to 100 students in a series of lectures and then those students passing a test. That is not education. That is a way that

1 To Stop Cheats, Colleges Learn Their Trickery http://www.nytimes.com/2010/07/06/education/06cheat.html?_r=0

universities can have 50,000 students while only hiring 2,000 professors, a model that really doesn't work for the students at all. Listening and regurgitating is not education. Suppose we actually tried to teach every student to think for themselves. Wouldn't we have to individually assess their actual thinking, by engaging them in a real conversation, to see if they can think clearly?

Second, and way more important here, is the plain fact that for the most part, universities aren't teaching students to actually do anything. They are teaching them to write papers about what they know, which is very different from actually doing something. You can't cheat in an engineering class if your job is to build a plane that flies and the professor's job is to watch it fly. You can't cheat in a music class if your job is to play the piano and the teacher's job is to listen to you. You can only cheat if your job is read and write and the professor's job is to grade essays as fast as he can.

As long as doing is subjugated to a secondary role in education, cheating will occur regularly. As long as being educated means being able to write an academic essay or being able to fill in dots on a multiple choice test, students should cheat. They are being cheated of an education and they know it, so they should cheat in response.

This is all a silly game and all students know it. What they have to do to get a degree is their question. No one is really providing them with an education. Professors can claim that they are teaching students to think, but they are more typically teaching them how to look at the world in the way the professor looks at it.

Perhaps it is time to start producing people who can do things and to stop worrying about students ripping off essays from the Internet. The simple solution is to stop having them write essays – but then someone might have to actually teach someone to do something and then watch and see if they can do it. That thought is horrifying to universities because it implies a different economic basis for the university – one not based on research contracts – as well as a de-emphasis on academic research for students who will never do it as adults.

When professors stop cheating students of an education, perhaps students will stop cheating as well.

As an aside, in 35 years as a professor I never once assigned a research essay or gave a multiple choice test. I did, however ask students to think and write about things that had no right answer. And I asked them to build things. I actually expected them to think.

Why do we teach what we teach?

When I was at Yale I occasionally did some freshman advising. I once had to advise a student who had just arrived from the Los Angeles ghetto as to what courses to take. I said to take what interested him, since there were very few requirements. That was not what he wanted to hear. He asked what I had taken as a freshman. I had attended Carnegie Tech 20 years earlier than the date of this advising session. There were no choices. You took calculus, physics, chemistry, western civilization, and English literature. No exceptions. He said that he would take those courses. I said that was absurd. He said that I was very successful and that that was how I had started, so that was what he would do. And that is what he did.

I was reminded of this story when I had a little encounter with the Obama administration last week. They are about to propose spending hundreds of millions of dollars on education to ensure that we do a better job of teaching the curriculum that has been in place since 1892. Re-examining what is taught and why it is taught will not be considered, because they are worried about class warfare. They don't want people saying that a new education system will be different from the one that got them to be the successes that they all are.

If Secretary of Education Arne Duncan and President Obama (and *New York Times* columnist Nick Kristof) got where they are by taking algebra and physics, then we can't take that away from the next generation of students.

This argument is so stupid it is hard to know where to begin to counter it. Let me just say that not all my friends from Carnegie Tech, nor from Stuyvesant High School for that matter, have become great successes. And those who have succeeded did so in spite of the mind-numbing "study for the test fact retrieval system" still in place in our schools.

I learned to think from my father, not from my school. Most successful people fare well because of good parenting and good genetics, not good schooling. We will always have winners in any system that is in place. School should help people live happier, more productive lives. It should not be about winning the competition. Of course, that is exactly what school is about now. Why do we believe that what is needed is to desperately pour money into an absurd system?

Why students major in history and not science

My 25 year-old niece offered to drive a friend of mine to the airport. As she was leaving, she returned to get a bottle of water. She said she needed to stay hydrated. I asked her if she was planning on taking the long route through the Gobi Desert. She seemed confused. I told her to look up the eight-glasses-of-water-a-day myth on the web. She is a young lady who fights big corporations with her every breath, yet she had bought into the bottled water companies' campaigns to make everyone carry water with them at all times. She was truly astonished to find out she was being manipulated.

I had less success with two other college students in recent weeks, who had both decided to be history majors because "history teaches you everything." While I suspect that there are way too many history majors for the available historian jobs out there, I have nothing against them, or against history. But I couldn't help but note that these kids had been sold history in the same way that my niece had been sold water. Every liberal arts college is desperately trying to stay relevant by selling the advantages of majoring in history or English to a group of young minds who have no idea how to make these decisions. The sellers look askance at practicality and tout students into their fields because if they don't, their departments would cease to exist. If no one majored in history there would be no history departments, except at the most elite and wealthy universities.

Would this be a bad thing? It is easy to assume that this would be a terrible thing. We assume this because we see universities as repositories of scholarship and wisdom. If that is indeed what they all are, all would be fine. But they are primarily places where young people start the rest of their lives. I asked one of these students what she loved about history and she replied that she was only really excited about astronomy. I asked why she wasn't majoring in that (she attends a university where she should do exactly that) and she replied "what could I do with that, discover another planet?"

There was no convincing her that the path she had chosen for herself was nuts. She planned to go to law school next – because it helps you think, she said – and she does not intend to be a lawyer.

Universities are doing students a disservice by perpetuating wrong ideas about what is worth studying. These study areas are really mostly intended to keep their most irrelevant faculty member employed. Science has been marketed badly and history has been marketed well. Business is marketed (and taught) terribly. Medicine is made so annoying to study in college that we have fewer doctors than we need. But we have plenty of history majors.

Maybe it would be good if universities stopped looking out for their own needs and started helping students make decisions that are right for them and their actual areas of interest. Of course this won't happen. Faculties run universities and faculties are always primarily interested in maintaining the status quo.

Why don't we encourage schools to adapt to kids rather than the other way around?

A couple of days ago I had a conversation with my daughter-in-law that was mostly about her son Max (my grandson). Most conversations with her (and with my son) are about this particular child because he is a handful. He is very bright, very verbal, and (today) generally obsessed with street maps (in fact, I just bought him a wall-size one for his birthday – it is what he asked me for). He doesn't do what he is told, he zones out, and he isn't good with other kids because they usually don't share his current obsession. His parents have worked hard to find a school that can handle him and, of course, they have had him diagnosed.

My subject here is an offhand comment from my daughter-in-law during our conversation. She said that a lot of the kids she knows of his generation have various issues and the reason is probably environmental. I assume it is true that where she lives, many kids have been diagnosed with some kind of disorder. For all I know, the reason really is environmental. I don't have much knowledge about how our current environment might be causing weird behavior disorders in kids.

But this is what I do know. Had Attention Deficit Hyperactivity Disorder (ADHD) been a diagnosis in the 50s, I would have been diagnosed with it, and my mother, who believed in doctors, would have had me drugged. Since there was no such diagnosis in those days, my mother would just show up at school and say, "he's bored, try giving something more interesting to do." She once even suggested to my third-grade teacher that I would be better off sweeping the floors than trying to put up with whatever they were teaching at that moment.

I also had a conversation with my daughter's son Milo this weekend (it was his ninth birthday). I asked him about school and he replied, as he usually does, "boring." I asked him what was boring now and he said they were learning about Indians again, just like they did in the second grade (he is now in the fourth), and that he was tired of it. I suggested that he ask the teacher why it was ok for us to have murdered millions of them, as that would stimulate good discussion, but no one had ever mentioned to him what became of these Indians he is learning about.

I am writing about this because the *New York Times* published a phenomenally important article yesterday.[1] Everyone interested in education should read it. Here is some of it:

> **A Natural Fix for ADHD**
> Recent neuroscience research shows that people with ADHD are actually hard-wired for novelty-seeking – a trait that had, until relatively recently, a distinct evolutionary advantage. Compared with the rest of us, they have sluggish and underfed brain reward circuits, so much of everyday life feels routine and under-stimulating. To compensate, they are drawn to new and exciting experiences and get famously impatient and restless with the regimented structure that characterizes our modern world. In short, people with ADHD may not have a disease, so much as a set of behavioral traits that don't match the expectations of our contemporary culture.
>
> From the standpoint of teachers, parents and the world at large, the problem with people with ADHD looks like a lack of focus and attention and impulsive behavior. But if you have the "illness," the real problem is that, to your brain, the world that you live in essentially feels not very interesting.
>
> One of my patients, a young woman in her early 20s, is prototypical. "I've been on Adderall for years to help me focus," she told me at our first meeting. Before taking Adderall, she found sitting in lectures unendurable and would lose her concentration within minutes. Like many people with ADHD, she hankered for exciting and varied experiences and also resorted to alcohol to relieve boredom. But when something was new and stimulating, she had laserlike focus. I knew that she loved painting and asked her how long she could maintain her interest in her art. "No problem. I can paint for hours at a stretch."

Why are so many kids being diagnosed with problems these days? Here are three answers:

- Drug companies have drugs they want to sell and they push these diagnoses. More illness = more money for them.
- Therapists have therapies they want to sell. More problem kids = more therapies to sell.
- And of course, the last and biggest. It is very difficult to get a group of kids to want to hear about Indians, or pollution, or math, or "science." But the schools insist on teaching things that most kids aren't interested in, and they are lots of kids in a class. The teacher can't put up with all these individuals who want to do what they want to do and are not interested in what they want to teach. So kids need to learn to sit down and shut up. Milo is a compliant kid so, although he is bored, he does sit down and shut up. Max (like me) would never sit down and shut up if he was bored or had something more interesting that excited him, to talk about or do.

1 http://www.nytimes.com/2014/11/02/opinion/sunday/a-natural-fix-for-adhd.html

The school's job is to excite kids about what is out there in the world and let them have a go at it. (Or that ought to be the school's job.) Instead, schools have taken on the job of babbling on about whatever the official (and out of date and irrelevant) curriculum happens to be. They have decided that kids who would rather be doing something else will not be allowed to do so.

Now this might have been the only possibility in a world of "mass education" and giant school buildings that look like prisons (not randomly). But today we have the Internet and mentors available online, and experiences that can be individualized.

Why don't we encourage schools to adapt to kids rather than the other way around?

I always told my children (and now my grandchildren) that when you don't understand why something is happening the answer is usually "money." How is the answer money here?

Drug companies are making a lot of money on ADHD drugs. Doctors make money on prescribing these drugs. Testing companies are making a lot of money on making sure kids sit down, shut up, and take the test.

The goal seems to be a drugged kid who has memorized the quadratic formula and is having no fun at all.

Why do we give lectures? Why does anyone attend them?

I recently found myself in the unusual position (for me anyway) of being a tourist in Brazil. For various reasons, I was on boats, buses, and other vehicles, on which I found myself being lectured at.

This was a bit ironic because as my readers know, I hate lectures, which is also ironic, because I am a frequent lecturer at meetings of one sort or another.

I found myself wondering why people love to give lectures so much, and why I seemed to be the only one irritated by having to listen to them. (One was given beneath a tree, so I walked away, but no one else did, and one was during a walking tour of a winery, which I left, but again no one else did.) Now, I understand why no one left the buses or the boats, but I certainly wanted to. On one boat ride (which I had thought was just a trip around the harbor), the man giving the lecture mentioned at least eight times that there were (fill in the number) states that comprise Brazil. I had no idea why he was telling us this, and, obviously, I have no idea what the number is (I am guessing between five and 50). I don't care now any more than I did then.

Why was he telling us this fact once, much less eight different times?

There is something about lectures that is fascinating to me because while I hate listening to them, I love giving them. In fact, it seems like most lecturers love giving them, so my question is why anyone listens.

As a professor (but one who did not lecture) I understand that students are there because they have to be, and for the most part they aren't listening much. But, I have noticed that most people will not admit this about themselves. When I ask people to try and remember a lecture they heard, they usually say they can and then say a sentence or two about one that they happen to recall. But the average educated person has heard hundreds of lectures, and usually cannot even remember what the subjects were or who the speakers were after a while.

So my question remains. People voluntarily submit themselves to this and they do think they learned something. Why do they do it?

Here are my best guesses as to why we give lectures and why people seem to want to attend them.

5 reasons why people give lectures

1. Everyone is looking at the lecturer and the lecturer is performing. People love performing in front of an audience.
2. A lecturer feels as if he or she is the smartest person in the room while lecturing. Everyone is paying rapt attention (they think), so they must be very smart and very important. People like being the smartest person in the room. Even the boat guy felt he knew more about Brazil than anyone else on the boat, so he was sure he must be very wise indeed.
3. The lecturer feels that he or she is saving time. If the lecturer can convey lots of information in an hour, think of the time the audience and the lecturer are saving by putting everything in one neat place.
4. The lecturer is also saving money. Instead of having a conversation with each member of the audience, he or she can talk to everyone at once. This makes university education very cost effective and does the same for corporate training. One person and five hundred listeners makes great economic sense.
5. A lecturer, not this one of course, believes that facts are the currency of education. The more facts he or she can provide, the better off everyone's life will be. If he or she could only talk faster, think how many more facts could be provided. The providing of facts must be thought of as being very important, even if one of those facts is the number of states in Brazil.

5 reasons why people listen to lectures

1. Everyone likes watching a performance. People listen to the State of the Union address to see the performance. People attend a keynote lecture at a meeting to see the performance. After more than 40 years of giving lectures, I have come to believe that most people haven't much of an idea what I am talking about and they don't much care, but they like when I make them laugh and they like when they can come up and talk (or argue) with me later.
2. People like feeling that they are smarter than the guy who is supposed to be the smartest person in the room. They get to tell their companions that the speaker was a dope, or make fun of something he did. They like feeling superior to the guy who clearly thinks he is the smartest person in the room.
3. People attend lectures because they are saving time. They get all the stuff they need in one place in one hour and then later they can "explore more deeply" if they want to. This is a nice myth anyway. I am not sure that much "exploring more deeply" actually happens, but it is nice to think that it does.
4. The attendee is spending money, not saving it. The lecture usually costs something one way or the other. But typically mom and dad, or the government, or the company, is paying for it so they don't care.

5. The listener agrees that facts are the currency of education. They like facts. They like them because they can pop them into a conversation at a cocktail party and seem erudite. (My wife heard these same tourist lectures. She is the opposite of me. She got all A's in school and was actually listening to the boat man. I asked her, while I was writing this, how many states there are in Brazil. She said 21, she guessed. I looked it up after she answered. There are 26. Later when I told her what I was writing, she said "oh it's 26.") But, even the good students don't really care much about the facts. They may say they are important but they know they are not (unless of course there is a test, in which case they are important for the test).

So, why do we have lectures? Because we always did. We are all just used to it, and no one wants to change this.

I will end with a quote from Max Sonderby. Max was the teacher assistant in the first learn-by-doing mentored simulation based masters degree program we rolled out at Carnegie Mellon's Silicon Valley Campus, in 2002. The year before, he had finished a typical masters degree at Carnegie Mellon using the classroom-based approach to education:

> I am almost jealous, in a way. I see that they are gaining skills more readily than I gained them in the program which I attended in Pittsburgh on Carnegie Mellon's campus. They get exposure to things that we just talked about in a lecture hall. They are actually doing it, implementing, building software, putting designs into practice, whereas we mostly just did homework and talked about it in a lecture hall. I am jealous in that respect, but it's also a lot more work, but that work definitely pays off for the student.

Max was right. Lecturing is a lot less work for everyone. We still have lectures for one main reason: they are the lazy person's approach to education. Both lectures and listeners agree that neither of them wants to do much work. Real work, and real doing, and real conversation, is all that matters for learning. But education is really not about learning.

Why our education system is the way it is: The Tillman Story

About the last thing I am likely to do in this space is to write about a movie. But, as it happened, I chanced upon a movie on TV in which I had no interest. Yet it had an impact on me anyway. The movie was *The Tillman Story*, which would mean nothing to non-U.S. people, and perhaps very little to many in the U.S. Pat Tillman was a U.S. football star who suddenly left the National Football League and his millions of dollars of salary to enlist to fight in Iraq after 2001.

The politicians in Washington loved this story, as it justified the "all-American hero fighting for his country" story that Bush and his cronies were trying to sell at the time. They played up the story in all the media. Tillman was killed in Afghanistan after some years and Bush and his buddies were busy touting the "our hero died for his country" line they love so much. The problem was that after some investigation on the part of Tillman's family, it seems he wasn't killed while fighting the enemy. Instead, he was killed by U.S. troops who just seemed to be having fun shooting anything that moved one day.

The Tillman Story details how the family uncovered the cover-up that the Army had created to obscure what really happened. The movie is unkind to the Army, but, as someone who has worked with the Army for a long time, I was skeptical that the Army would be that involved in telling such an elaborate lie. Eventually the movie points the finger at Donald Rumsfeld, who appears to have been calling the shots, and makes it clear that George W. Bush would have had to have been involved as well.

My first thought was that it says something that they were allowed to make this movie at all. A repressive government doesn't let you make anti-government movies. The U.S. government may have many faults, but freedom of speech still exists here.

But then, my thoughts turned to the real subjects that always interest me, which are stories, and the general stupidity of the American public.

The lengths to which Bush and friends went to tell the Tillman story that they wanted to tell, and to cover up the real story, are well-documented in this film. Why? Why lie, cover up, misinform, hush people up, manipulate the media, and otherwise be hysterical about the fact that a soldier was killed by his own troops? This happens all the time. It is called the fog of war.

The answer is that stories matter. Politicians love to tell stories and the stories they tell often have little relation to the truth. They get away with this because stories are simple and easy to understand. The truth is often much more complex.

This points to one reason why politicians all seem to agree on testing and generally making our education system about memorization of facts (otherwise known as "official stories"). What we want students to learn is what the true stories are. We want them to know the facts about George Washington and Abraham Lincoln and Pat Tillman. We really don't care if those facts are true. In all nations, the job of education is the telling of official government-approved stories, about everything from history to economics to how to be a success and why to fight for your country. No one cares about the truth all that much. They just care about having good stories to tell.

We are all susceptible to a good story – that is why we like to watch movies in the first place. It is not just poorly-educated people who like simple stories. We all do. It is part of being human. But how do we learn to determine if a story is true?

We wouldn't have known the truth about Pat Tillman if it hadn't been for his family being smarter than your average family and really wanting to know what happened. They were capable of separating truth from fiction. But this is a skill that we are more or less explicitly taught not to do in our schools.

What can be done? Ask students to think instead of memorize? I have been saying that for years, but, no surprise, no government official is ever on my side on that one. They like being able to tell simple stories that remain unexamined by their listeners.

Why not use the principles of prison reform to help schools?

Last week I met a man who was interested in investing millions of dollars in fixing education. I was happy to meet with him. But within minutes, it became clear that his idea of fixing education was very different from mine.

This man was concerned with making the system more efficient. He had no concerns about making school fun, interesting or useful in later life, nor was he concerned about offering more choices, forcing fewer requirements, or getting rid of tests. No, quite the opposite. He wanted to test everyone and everything all the time. He wanted to use computers to efficiently deliver tests, grade tests, evaluate teachers, deliver materials, and so on. Here is a summary of what he was discussing:

- As with any government program, the public school system must be transparent and include performance measures that hold it accountable for its results.
- Colleges, along with taxpayers and the public, are among the key "consumers" of the school system; their interests should be prioritized when determining appropriate education goals.
- The education system should emphasize personal responsibility, work, following the rules, and remediation while supervising students.
- An education system works to reform underperforming teachers and students and put students on the right path.
- Because incentives affect human behavior, policies for both students and teachers must align incentives with the goals of economic growth, good citizenship, increased college attendance, and cost-effectiveness, thereby moving from a system that grows when it fails to one that rewards results.
- School reform should not be used to grow government and undermine economic freedom.

He never actually said any of this. He reminded me so much of someone who was interested in running an efficient prison. To compile the list above, I looked up "prison reform principles" and adapted what a set of prominent conservatives had written on the subject.

I think school reform and prison reform have a lot in common. As long as we think of schools as a kind of prison, where students and teachers do what they are told, when they are told, with no freedom at all and constant assessment, maybe we should adopt prison reform ideas wholesale and simply forget about caring about children or helping them think clearly. After all, no one wants a prisoner to think.

Why educate the elite? A lesson from Yale and George Bush.

I have been doing press conferences lately on behalf of Experimental Teaching Online (XTOL) Europe, which is offering our learn-by-doing courses. You get some strange questions whenever you talk to the press, but today I got one that was not only odd but set me thinking about something new. In Spain, which is where I am as I write this, a lot has been written about me, so reporters come armed with what they read in some other press coverage. The question that set me off was this:

"You say school was designed to educate the elite classes and you don't care about them. Why don't you care about educating the elite?"

Gee, I had never thought about it that way. I have said that Harvard, Yale, etc. can get away with letting people major in History, English, Classics etc., because historically their graduates were the sons of the rich and ruling class and no one expected those graduates to get a job. While that is not exactly true today, believe me, Yale and Harvard haven't changed that much. But truly, I have never been concerned with changing those institutions. I loved my time at Yale, and kids don't exactly come away ruined from spending four years there. My complaint has always been that all the other universities have copied this elitist model. In fact, in Spain, one hears constantly about the value of studying literature, in a society with massive unemployment amongst youth precisely because so many students have studied relatively useless subjects.

I don't worry about the education of the elite classes. I worry about average Joe who can't think clearly and whose skills have not been enhanced by school.

But this question made me think about the elites. What should we be teaching them? Harvard and Yale keep graduating future presidents, Supreme Court justices, governors, and business leaders. What should we be teaching them?

Amazingly, literature, etc. still doesn't come to mind. Yes, of course we would not like our President to say "who?" when Dickens is mentioned or to say "what?" when the Peloponnesian War is mentioned.

But what should we be teaching them? Oddly, George Bush, famous Yale graduate (pick either one), comes to mind. Was the problem that these men weren't well-versed in the classics?

Here is a thought. Neither seems to know much about average Joe's concerns, what it's like to work for a living, what the average schoolroom is like, or, for that matter, how the economy works or how to govern.

I have to admit, a little more knowledge of history wouldn't have hurt either of these men. But what history should they have known? I would have hoped they might have known more about Arab society or the history of Iraq or Afghanistan. Why do I feel confident that they did not study the Middle East at Yale?

I used to be on U.S. Army's subcommittee on distance education. As part of that I had the occasion to attend the Army War College for a couple of days. I was with a group of majors and colonels who were learning about the history of Islamic revolution. I doubt that the Bushes took that course at Yale either.

So, yes I am worried about the education of elites. I think they should learn how the other half lives (maybe by living with them). I think they should learn how to govern – maybe by running a smaller country first (I am joking – sort of). I think they should learn real economics and how to diagnose a problem and how to say things that are more than sound bites. They should learn how to lie less, and how to manage on less.

I realize this will never happen, but I think it's fun to think about. The French, by the way, do have a school for training future political leaders, but I get the idea it hasn't worked out that well.

The top ten mistakes in education: twenty years later

It has been 20 years since I wrote about the top ten mistakes in education for my book *Engines for Education*.[1]

Twenty years have passed. Surely my writing about them, and other people's re-posting and writing about them have had a big effect on education. Let's revisit and update them one by one:

Mistake #1: Schools act as if learning can be disassociated from doing.
Yes, things have changed. They are worse. The latest horror are MOOCs (Massive Open Online Courses), which is just more talking, and insists that education means knowledge transfer and knowledge can be acquired by listening.

Mistake #2: Schools believe the job of assessment is part of their natural role.
Things have really changed here. They are much worse. Before, there were just lots of bad tests. Now there are tests at every grade. Tests to get ready for the test. And now, teacher evaluations are based on the tests.

Mistake #3: Schools believe they have an obligation to create standard curricula.
Wow! This one has gotten even worse than the others. Now, it isn't schools that create standard curricula, it is Bill Gates, Common Core, the U.S. Department of Education, and every state Department of Education. We sure fixed that one.

Mistake #4: Teachers believe they ought to tell students what they think it is important to know.
I am not sure about this one. I don't think teachers think much of anything anymore, other than how to survive in a system where they are not valued and teaching doesn't matter except with respect to test scores.

Mistake #5: Schools believe instruction can be independent of motivation for actual use.
No change. There is still no use for algebra, physics formulae, or random knowledge about history or literature. No use for anything taught in school actually, after reading, writing, and arithmetic.

1 http://www.engines4ed.org/hyperbook/nodes/NODE-283-pg.html

Mistake #6: Schools believe studying is an important part of learning.
No change.

Mistake #7: Schools believe that grading according to age group is an intrinsic part of the organization of a school.
No change.

Mistake #8: Schools believe children will accomplish things only by having grades to strive for.
No change.

Mistake #9: Schools believe discipline is an inherent part of learning.
Perhaps this has changed. There seems to be a lot less discipline.

Mistake #10: Schools believe students have a basic interest in learning whatever it is schools decide to teach to them.
Nah. No one believes that anymore.

I am not only one loudly talking into the wind. There are lots of people who agree with me and say things similar to what I say.

Is anyone listening?

Sure. Parents are noticing how stupid the tests are and how stupid Common Core is. The kids are noticing, now more than ever. The teachers are upset.

Is anyone listening to them? No. There is big money at stake in keeping things as they are.

Well, that's the report from 20 years on the front lines. We shall not retreat, but victory looks to be far away.

Chapter 2: Teaching the Wrong Things

Wrong problem, wrong solution

Math and Science, oh my. What will we do? We don't produce enough students interested in math and science. Something must be done. I hear this refrain so often my head hurts.

First, my credentials: I was a math major in college. I got 98 on every math Regent's test offered (I lived in New York, where testing ruled the world in the 1950s too), and my mother always asked where the other two points went. I grew up to be a computer science professor. I am not a math-phobe. But nor am I a math proponent. I never used math in my professional life, never ever.

I start any discussion on education by asking if the person I am talking with knows the quadratic formula. One out of 100 knows it. The last few people I asked included the head of a major testing service, the secretary of education of a U.S. state, various state legislators, and 200 high-school principals. So why do we teach this obviously useless piece of information to every student in the world? Because math is important, of course.

Really? Show me the evidence.

As a person who did graduate admissions for 30 years at three of the top ten universities in the country, I know what this hysteria is actually about. Nearly all applicants to graduate computer science programs are foreign nationals (and this is true in most fields of engineering and science). We wonder why American kids aren't interested in these fields – which is a reasonable enough question. But then we have come up with an extraordinary answer.

What we say is that we must teach math and science better in high school. There are now so many programs aimed at achieving this, it makes my head spin. Here are reasons why this is simply the wrong answer.

Do we really believe that the reason there are so many foreign applicants to U.S. graduate programs is that they teach math and science better in other countries? China and India provide most of the applicants. They also have most of the people. And many of those people will do anything to live in the U.S. So they cram math down their own throats, knowing it is a ticket to America. Very few of these applicants are coming from Germany, Sweden, France, or Italy. Is this because they teach math badly there or is it because those people aren't desperate to move to the U.S.?

In the U.S., students are not desperate to move to the U.S. So when you suggest that they numb themselves with formulas and equations, they refuse to do so. The right answer would be to actually make math and science interesting, but with those awful tests as the ultimate arbiter of success, this is very difficult to do.

No change in education will ever happen in the U.S. until the testing mentality is done away with. No average high-functioning adult could pass them, so why make kids do them? It makes no sense. What also makes no sense is the idea that math and science are important subjects. You can live a happy life without ever having taken a physics course or knowing what a logarithm is.

On the other hand, being able to reason on the basis of evidence actually is important. Thinking rationally and logically is important. Knowing how to function in a world that includes new technology and all kinds of health issues is important. Knowing how things work and being able to fix them, and perhaps design them, is important.

Let's get serious. *We don't need more math and science. We need more people who can think.*

We need to teach job skills, people skills, and reasoning skills. And we need to make education exciting and interesting. We need performance tests, not competence tests. If we did all this, we would get more Americans interested in math and science, because we would get more Americans actually interested in being in school.

In defense of what doesn't work

I love the defenders of the faith. Attack teaching mathematics, and you are told that kids need to learn to add. Mention that history is a waste of time, and random facts are mentioned that students don't know and everyone is appalled. Say that English literature is a ridiculous subject for high school students, and the fact that kids can't write or speak properly is offered as defense. Mention why books are no longer important in today's world, and librarians throw a fit. Suggest that science is not about memorizing formulae, and people who have no idea what scientists actually do all day tell you that scientists need to teach everyone more facts about how the world works.

Trying to get people's arms around the real problem in education is not that easy.

The reason is you.

You all went to school so you are quite sure that what is taught in school is what should be taught in school only that we should teach it better.

This is how you figure out what should be taught in school: Ask successful adults what they do all day and check how often different skills show up.

- Calculus – not so much
- Literary analysis – not every day
- Physics formulae – never see them again after high school
- Libraries – can't you get free Internet there now?

This country needs to come to grips with the fact that the high school curriculum reflects a notion of how nineteenth-century scholars thought about how to produce more scholars like themselves.

Dropout rates of 50% in high school reflect the irrelevance of what is being taught. The kids know it, but the system, which is defended by nearly everyone associated with it, does not. Here are some obvious truths:

- To teach someone to reason, one does not have to teach about congruent triangles.
- To teach someone to write effectively, one does not have to ask them about themes in Shakespeare.
- To teach someone about daily economics, one does not have to teach about tariff acts.

- To teach someone to be a good citizen, one does not need to know about Lincoln or Washington, but about how to analyze for truth what current Presidents are saying.
- To teach someone to be employable, one does not have to teach nearly any subject required by colleges for admission.

Let's think again folks. Education is about teaching people how to live and how to make a living (to paraphrase John Adams). We have plenty of intellectuals. Feeding the colleges is not the priority of the modern day high school – making high-functioning citizens is.

The curriculum deciders

I was giving a talk that mentioned how the curriculum in the schools is outdated and irrelevant and needs to be thrown out. Of course, I got the usual questions. People defend their favorite courses but are willing to trash the ones they didn't like or do well in. And then there is my favorite question: "Who will decide what the new curriculum will be?"

There are two serious problems with that question. First, why must there be one curriculum? Second, why should anyone but the students decide what they should learn?

Of course I know that these problems are rarely mentioned. We just assume that there should be exactly one course of study in high school and that students should be told what they have to learn. These assumptions are so strongly held that any suggestion by me that school needs a redesign sparks the assumption that I want to dictate what the curriculum should be. I don't. What I do want to do is to design as many curricula as possible to allow as much choice as possible. Someone I know told me that his daughter was bored in school and that what she really wanted to learn about was set design. And why shouldn't there be a "set design" curriculum for those who want that?

Since curricula can now be delivered online – as can teaching – the old excuse "not enough demand" goes away.

Of course there is another objection – that such a curriculum wouldn't include the important stuff. Really? What is "the important stuff"? I am sure that one actually has to know a great deal about "the important stuff," whatever that is, in order to do set design or anything else that is part of the real world. The place to teach the important stuff is within a context of interest to the student where it would actually be used.

We need to understand that our unspoken assumptions about education are wrong. Every high school drop-out knows they are wrong.

Benito Juarez, Don Quixote, and Mexico City street kids

I have just returned from a long trip that included Mexico City and Riyadh among other places. Riyadh was certainly fascinating in many ways, but my mind keeps returning to Mexico. This is because the people who invited me to Mexico (Telefonica España, a Spanish telecommunications company) made a serious mistake – they invited me to visit a school.

Now this wasn't just any school. It was a school for street kids, those who sweep up, sell Chiclets gum, run errands, or do anything else they can to earn money. Telefonica had convinced the parents of these kids to let them go to school until noon, enabling them to work the rest of the day. Clearly Telefonica is trying to help, but the result isn't clear. They don't have the freedom to design the half-day they provide for these kids. The Mexican government, like all governments, still dictates the curriculum. So, the first lesson I observed was about what a wonderful man Benito Juarez was.

The picture is of the teacher asking the class questions about Benito Juarez. Notice that the children are all sitting in front of computers, but they are not using them, they are listening to a lecture. They were using them at an earlier point to play with a jigsaw puzzle that had a map of Mexico on it, which as far as I could tell was just about finding pieces that fit, and not really about Mexico at all. Later the students went into the yard to march and sing a song about Benito Juarez.

The next day, I gave a speech about "the role of the teacher in the twenty-first century," which was the theme of the two-day meeting.[1] Telefonica has been holding discussions about my speech in most of the Latin American countries and in Spain for the last few weeks.[2]

For years I have been pointing out the absurdity of the curriculum taught in schools. It is outdated and irrelevant to most children. But in Mexico, I saw a serious need for schools that help these children lead better lives, for which I have much empathy, but still the school authorities don't get it.

Benito Juarez, really? (I have the same point of view about George Washington, by the way). So, when these kids go back to the street (and quit school at 14, as I was told they all do) they will know about their national hero. They won't know how to get a good job, however.

We could be teaching them how to run a business, how to program a computer, how to raise a family, how to find ways to make money – really anything that might help them get out of poverty. But no, they sing about Juarez.

Now I wish this were just a problem in Mexico, but it is a problem anywhere and everywhere. As it happens, I speak in Spanish-speaking countries more often than I do in English-speaking ones, so I have learned that certain things I say will drive Spanish-speaking intellectuals mad, but will get them to think harder about what they are doing to kids.

I always say, for example, that they should stop teaching history. This remark is generally hated in Spain, but apparently I upset a few people in Mexico as well. But how accurate is the history we teach? Mexican history as I understand it is about the Spanish occupation and the mistreatment of the local population, some of which is still going on, as can be seen from the color of the children's faces in these pictures. The descendants of the Spanish are not selling Chiclets.

If you look at the Spanish text in the online debate held after my talk, you will see that the conversation focuses on several topics. One of these is about my remark that teaching Don Quixote is not necessarily the cleverest idea. Don Quixote is required reading in Spain and, as far as I know, in every other Spanish-speaking country. It is defended as a good way of learning about their culture. How events in Spain in 1605 (or really a novel set in 1605) are of value to Mexicans who really have no need to know about Spanish culture, I don't know. Many say that Don Quixote was the best novel

1 My speech can be found here: http://www.youtube.com/watch?v=klquall9HjM

2 Some of the basis of those discussions can be found here (in Spanish) http://blogs.educared.org/encuentro/2013/02/01/jornada-1-evento-en-mexico-un-debate-muy-politizado/

ever written. Maybe it was. I don't know. I have managed to live a good life without ever reading it, which is probably true of most Americans (and American intellectuals) as well.

The only relevant question for me is how these street kids can be helped to live better lives. I don't get how making them read Don Quixote and sing about Juarez will help. Further, I don't get why the Mexican government, and the Spanish-speaking intellectuals who argue with me about this, cannot comprehend that what they are doing is simply wrong.

It is all well and good to have a view of education that says we need to teach our culture and history and where we came from, but this idea flies in the face of reality. Spain isn't where most Mexican street kids came from. (A little-known fact here is that Spain is actually where my ancestors came from and this makes me not one bit more interested in reading Don Quixote.)

Culture and history are very nice for intellectuals. Let them have it if they want it. And maybe, through Telefonica's efforts, one of these kids will stay in school and become an intellectual. But is that really anyone's goal? We have plenty of intellectuals. What about the rest of the people? What about the kids who will, at best, live ordinary lives?

Why can't we teach them skills that will help them with their lives and why can't we start doing this now? Teach them to be healthy, productive, have some fun, think clearly, and be good to others. Teach them life skills.

The answer is that there is no government I know that actually has that goal. (Maybe Saudi Arabia – they hinted at it while I was there, but I don't know if they meant it.)

Governments design schools to make citizens who will behave themselves and who will not threaten the status quo. But the status quo isn't so good in Mexico (or in my country either, frankly). It is time we stop trying to turn everyone into an academic scholar and start trying to help students become functioning adults who can earn a living, raise their children, and get along with each other. Don Quixote won't help a bit in that kind of education and neither will Benito Juarez.

Stop pushing math in the name of reasoning

As I write this I am at sea, both literally and figuratively. I was just "getting away from it all" for a week, but now as the week comes to an end, I see I simply can't "get away from it" in any way. By "it" I mean the general absurdity of the nonsense we say and do about education.

Today I read an article about an Organisation for Economic Co-operation and Development (OECD) report.[3] The article talks about how terrible it is that Sweden's PISA scores are slipping, and quotes the U.K. education minister blaming the U.K.'s poor performance on the Labor party. Do we really need international math contests?

Apparently we do, because OECD's real mission seems to be to standardize teaching around the world. I have been loudly opposed to Common Core's attempts to do this in the U.S., an idea being pushed by Bill Gates for reasons best known to himself.

The standard canard that one hears in every report about low math scores is that math teaches reasoning and problem-solving skills and is critical for surviving in the twenty-first century. OECD says that their "mathematics test required creativity and problem-solving skills based on a deep understanding of mathematical concepts."[4] Uh huh. "Math teaches creativity and problem solving" has almost become a religious proverb.

I have this odd idea that one should have evidence for statements one makes. Especially when statements are made by large organizations that affect everyone.

Where is the evidence that math teaches problem solving and reasoning? It doesn't exist.

As an example, I will talk about myself for a minute. I was a math major in college. I liked math and was very good at it.

Now let's talk about how my math ability has helped me solve problems and reason in my own life. We all do many kinds of reasoning, but three things most of us need to reason about are relationships with other people, business/everyday decision-making, and decisions about our own health.

Let's start with the last one. I am getting older and health decisions come up again and again. I find that I am not particularly adept at making them. For one thing, I don't know enough. Also, doctors contradict each other, so I have to figure out who to trust. There are many issues I worry about and nearly as many answers about them from

3 OECD warns west on education gaps. http://www.nytimes.com/2013/12/09/world/asia/oecd-warns-west-on-education-gaps.html?src=rechp&_r=0

4 OECD's Programme for International Student Assessment (PISA) http://www.oecd.org/pisa/test/

various sources. Maybe I am better at making my medical decisions than others and maybe I am not, but my math ability has absolutely nothing to do with it. If only I were a better mathematician than surely I would be great at making the medical decisions I need to make? Does this sentence make sense to anyone?

Oh, but personal decision making, my math ability has surely helped there, right? Actually, I can't think of any area of my life that it has helped less. Love is not an equation. Nor is parenting, nor do relationships at work go well because you can do algebra.

Business? I suppose it depends on what kind of business you are in.

The voyage I am on as I write this has taken me to some very rich places and some very poor places. In one of poor places, a place I have been to many times, I met with a university president. He is worried about the education he is offering to his students, because there really are no jobs for them where he lives. We discussed teaching practical business and entrepreneurial skills. We did not discuss the need for more mathematics.

Later I visited a very wealthy place, a place where people who are rich have second and third homes. The other people there work for low wages to help rich people live easier lives. More mathematics would have helped the poor people there, I am sure. They could reason better and then... Ooops – they would still be stuck living where they live in the economic and cultural situation that exists there. But surely the rich people got there by reasoning so well because they learned mathematics.

This sounds so silly it is difficult to write it without laughing. Rich people become (or are born) rich for many reasons. Were they all good at math?

Just as I was asking myself this question, my ship passed by a private island that I recognized because it belongs to a friend of mine, and I had been there. Is my friend very good at mathematics? Yes. It turns out he is. Does he make money from being good at mathematics? Yes. It turns out he does. So how is my friend doing in other areas of his life and in other decisions he makes? To my mind, not so well, but it is not for me to judge.

Suffice to say, mathematics ability does not teach reasoning in general. Why don't we teach reasoning in general? Everyone agrees that it is very important. Maybe we don't know how.

But we do know how. The problem is that mathematics is easier to test. Reasoning would be more amorphous, there would be less certainty about right answers, and in fact there would be many possible answers. There is also a cultural component. Reasoning about how to fix a social or economic problem would be different in any given place because the answers would depend upon the many factors that make up that place. There are good places and bad places to build a luxury hotel, for example. While some simple mathematics would certainly be part of the decision-making process about such a business idea, the answer would depend upon many factors, most of which would be difficult to assess in a multiple choice test.

OECD has to get smarter. Pay attention to your own name. **Teach economic and cultural development**. Stop the nonsense about PISA scores and start thinking about what kids in different populations need to learn to do. Reason? Solve problems? Sure. Teach them to solve real problems, ones that exist in the environment in which they live. Forget the math problems.

Why do we let the general public decide what should be taught in school?

An article in *The Atlantic*, "How Alcohol Conquered Russia" caught my eye the other day.[5]

It contained the following paragraphs:

> "The Kremlin's own addiction to liquor revenues has overturned many efforts to wean Russians from the tipple," as Mark Lawrence Schrad wrote in the *The New York Times* last year. "Ivan the Terrible encouraged his subjects to drink their last kopecks away in state-owned taverns" to help pad the emperor's purse.
>
> "Before Mikhail Gorbachev rose to power in the 1980s, Soviet leaders welcomed alcohol sales as a source of state revenue and did not view heavy drinking as a significant social problem," as Critchlow put it. In 2010, Russia's finance minister, Aleksei L. Kudrin, explained that the best thing Russians can do to help, "the country's flaccid national economy was to smoke and drink more, thereby paying more in taxes."

This column is about education. What could this article have to do with education?

There is a cynicism in government that includes the idea that the people really aren't that important – the sentiment being, in this context, that "they die early, but at least they pay taxes." This isn't too far from "they are stupid and can't think clearly, but at least they pay taxes."

Do most governments want their citizens to be stupid? Let's examine some evidence.

Yesterday, there was an article in the *New York Times* titled "Creationists on Texas Panel for Biology Textbooks."[6] In summary, a committee of people with a religious agenda is deciding how biology should be taught in Texas.

In *Time* magazine, we see "Atheism Added to Irish School Curriculums: A new lesson plan will teach 16,000 Irish schoolchildren about atheism, agnosticism and humanism."[7]

I am reminded of the famous remark made by Mark Twain (about 100 years ago): "In the first place, God made idiots. That was for practice. Then he made school boards."

5 http://www.theatlantic.com/international/archive/2013/09/how-alcohol-conquered-russia/279965/

6 http://www.nytimes.com/2013/09/29/education/creationists-on-texas-panel-for-biology-textbooks.html?pagewanted=all

7 http://newsfeed.time.com/2013/09/30/atheism-added-to-irish-school-curriculums/

Why do we let the general public decide what should be taught in school? The answer is because we really don't care what is taught. By this I mean that the verb "to teach," when used in a school context, means to tell students what is true so they will believe it. What does it mean to teach atheism (in Ireland), for example? It means people will attempt to teach ideas that oppose Catholicism, which is what dominates the schools there. I am sorry, but did it ever cross anyone's mind that both sides are wrong?

Schools should not "teach" anything. Why not? Because what we really mean by "teach" is "indoctrinate." We want to tell students what to think. Little thought (but much lip service) is devoted to teaching them how to think. We want to "teach" students to be good citizens, to "teach" them about our history, to "teach" them math and science. These last ones are not exactly indoctrination. But what they actually are reminds me of the story I started with about drinking in Russia.

Math and science are meant to teach thinking (or so it is said). They could actually teach thinking, of course, but when the scientific questions are given to you, and the right answers are taught to you, science ceases to be about observation, experimentation, hypothesis creation and reasoning from evidence, and becomes memorization to get good scores on multiple choice tests.

How does this relate to Russia's drinking problem? Those who follow the rules and memorize everything they are told to memorize will probably turn out to be obedient, tax-paying citizens. It is all really the same idea.

Yesterday I was watching NFL football. The face of Sal Khan came on – I don't know how many times – talking about videos that Bill Gates and now Bank of America are backing. Why do they back these small lectures that are meant to get the current high school curriculum banged into students' heads? Why national TV ads? Because those in power want everyone to do what they say, memorize what they say to memorize, and avoid thinking hard about real issues. Cutely done math tutorials are the latest thinking about how technology can fix education. No one thinks about changing a curriculum that was written centuries ago.

You can't fix something that doesn't want to be fixed. You can help those who have bought into the system do math better and aid their chances of getting into Harvard. But what about the rest of the people? As long as they shut up and pay taxes, Bill Gates, Sal Khan, and now Bank of America, will be happy. No one really cares about the average Joe (or Ivan).

They never have.

Why textbooks suck

I take as my starting point a textbook written by a friend of mine, so certainly the best textbook ever written. Just kidding.

LIFE has chapters that start with a question. This is very good. Here are some examples of subchapter headings:

1.1 What is Biology?
1.2 How do biologists investigate life?
2.2 How do atoms bond to form molecules?
3.2 What are the chemical structures and functions of proteins?
4.5 How did eukaryotic cells originate?
7.1 How does glucose oxidation release chemical energy?
10.2 How do alleles interact?
12.4 How is RNA translated into proteins?
19.2 Is cell differentiation reversible?
26.6 How do prokaryotes affect their environment?
32.2 What is a protostome?
39.3 How do plants deal with climate extremes?
45.3 How do sensory systems detect mechanical forces?
51.5 How does the mammalian kidney produce concentrated urine?

The textbook contains very pretty pictures and diagrams and lots of self-quizzes. It does not, however, give me a reason to want to know the answers to any of the questions posed by the subchapters. It presumes that the simple fact that a student has signed up for a biology course is sufficient grounds to decide that the students have these questions. Or, possibly, it assumes that the professors, in giving a lecture, have raised these questions in the students' minds.

In my experience as a student, my main questions were usually "how much more of this stuff do I have to read?" (the book is over 1200 pages), and "could I just skip it and get by?"

My experience as a professor was that the most prevalent student question was "what exactly will be on the test?"

My experience as a department chair was that lazy professors answered that last question by copying the self-study questions in the textbook.

We would all agree that a good course motivates students to naturally ask the questions that a textbook answers, so they can consult the textbook in their moment of need.

There are two assumptions that need to be made here:

1. That professors know how to raise these questions in students' minds in a natural way.
2. That the book is well enough organized to make finding the answers trivial.

As a student I always wondered why professors didn't just hand out the textbook and say "read it and there will be a test in three months," since their lectures were usually irrelevant. Sometimes it was the textbook that was irrelevant. The students needed to quickly figure out which one wasn't going to be on the test.

But if real learning is not a conscious process, as some (i.e. me) insist, then textbooks could only be ancillary to what a student is doing. But what are students doing? They are sitting and listening, which is a conscious process.

Now let's imagine a world in which students are doing something. And, let's assume that they want to do what they are doing and are excited by it. And let's also assume that they know what "success at doing something" looks like. Then the textbook, in that case, would look exactly the same as it had before, except that chapters would be indexed to the goals of the students and the tasks that they were pursuing.

In other words, a textbook is like a collection of answers to questions that no one ever has. Making sure students are actually asking the questions that one might want them to ask would mean making sure they were pursuing tasks that naturally raised those questions.

This is the role of online education. It can create the environment for an answer to be relevant to the pursuit of a goal, by creating scenarios in which those questions naturally arise. This scheme, however, eliminates an important part of the school experience.

It eliminates the instructor. In the world I am envisioning there are no more lectures, just mentors who help students when they are stuck.

Instructors would object. Students would not.

Sacred truths

When we think about education, we typically imagine that its purpose is to teach students how to think. This is a very nice idea that has very little basis in fact. School was never meant to teach thinking. Schools have their origins in religious education. It is well to remember that Harvard and Yale started off as divinity schools and that until recent times, nearly all universities required religious training as part of the curriculum.

If we think about religious education for a moment, it doesn't take long to realize that, pretty much regardless of the religion, religion is about telling people what to believe and is not about questioning those beliefs. All religions know the truth and all religions attempt to dictate that truth to their followers. Most religions also run schools. No one criticizes them for this.

Our public schools have adopted the basic tenets of religious schooling:

- There is a truth that cannot be questioned
- There is no real choice in what a student learns
- You can be punished for failure to attend school
- You will learn by being told
- There are official sacred books that everyone must know

What are the sacred books of our schools? Shakespeare, Dickens, *To Kill a Mockingbird*, and *The Great Gatsby* are some of them.

What truths cannot be questioned? Algebra teaches you to think. You must know science to have a job in the twenty-first century. All of U.S. history is as it is depicted in textbooks.

Over the years I have said that all schooling needs to be re-thought. What we teach now was determined in the nineteenth century, and was meant to turn the few people who actually attended school at that time into intellectuals. When I say "get rid of all of it," the response is usually: *you are right about subject, X but subject Y is sacred.*

Sure, let's get rid of balancing chemical equations but we can never get rid of history.

Sure, let's get rid of algebra but literature is very important.

We can't get rid of science because it is important for knowledge workers.

This is what religion sounds like.

Curiously, school is still teaching religion. But now, the religion is about the sacred texts in which one finds the quadratic formula, or SP3 binding (you can look it up if you like), or what Julius Caesar said to Brutus. It is a very odd religion – one in which Shakespeare, Archimedes, Fermat, Descartes, Melville, and George Washington are gods.

None of this teaches children to think any more than the catechism teaches children to think. School ought to be a place where open minds can explore. This doesn't happen, because schools are simply the places where modern-day religious instruction can be found.

An honest back to school message to students

Every fall, many people look at my blog post entitled "Back-to-School Message to High School Students Who Hate High School." Curiously, it is clear that most of these people are principals and teachers who were preparing a "back-to-school message" for returning students on their first day back. My post is not useful for these speech-givers, since it explains why the subjects students are learning are out-of-date and irrelevant. But Google just finds "back-to-school message" and offers them my post no matter what else they may actually have asked.

Having said that, I propose a back-to-school message that teachers and principals can give to returning students that is actually in tune with the reality of our times. Here it is:

> Welcome back students. I know most of you wish you were still on summer vacation and that very few of you are happy to be back attending classes (although you may be happy to be back to the social and athletic aspects of school). But here you are, and there is nothing any of you can do about it. The government requires that you sit here, bored or not.
>
> That said, what should you be thinking about now that you are back? Whether you want to think about it or not, there will be lots of tests. The testing companies have sold the government on the importance of testing, so this year there will be more tests than last year.
>
> But you should not be particularly worried about these tests. Why not? Because they don't matter. Oh, they matter to the teachers whose salaries may depend on them, and they matter to the principal whose job may depend on them, and they matter to the school boards who will close, open, repair, and otherwise obsess about which schools are doing better than which other schools. And they matter to parents who don't understand what the testing is all about but who are sure they want their kids in the best schools (with the best average test scores), and of course those parents want their kids to have the best test scores (so they can brag to others about it).
>
> But test scores don't really matter at all to you. Why not? Here is why:
>
> - Nothing bad happens to you if you get bad scores. Oh, people may say stuff to make you feel bad, but nothing bad will happen to you in any way.
> - Something bad does happen if you obsess about getting good test scores. You will spend time memorizing stuff that will never matter to you, and you will be good at test taking, but you will not get any better at learning, thinking, creativity, or having fun.

- No matter what, you will go to college if you want to. There are 4,000 colleges in the U.S. Most of them will take anyone who applies.
- You may not get into the special school that takes only kids who have great test scores, but think about what those schools are like. Full of kids who study all day and do exactly what they are told to do in order to get ahead. That's not who you want to hang out with.

So what should you do this year? Most of all, have fun. Follow your passions, whether those are in school subjects or things that have nothing to do with school. Make friends. Have a social life. Learn to communicate better. Learn to get along with people better. And, learn to think better. Challenge yourself with things that are difficult to do and try, try again to accomplish them.

Oh, and stay off your phone. There isn't much to be learned from staring at someone else's party pictures.

NO to subjects and NO to requirements

I have been spending a great deal of time in Europe lately, where the talk is about what to do about the awful governments that countries like Italy, Greece and Spain seem to be saddled with. (I am not saying the U.S. government is any better – maybe it is even worse – I am simply reporting what I am hearing.)

In the course of one of these conversations, the talk turned to education, as it tends to do when I am around. The suggestion was made that schools should require students to learn about how government works, or maybe how it should work, in order to help citizens make better choices about who governs them and to be better at governing when they are actually part of the government.

I replied that this was a fine idea, especially if we let students run simulated governments rather than simply learning political theory. Feeling emboldened, a woman who had raised a family and who, I think, felt that she hadn't done such a good job, asked if maybe some courses in child raising shouldn't also be required.

I certainly agree with this as well. When I was building an online continuing education program for Columbia University, I tried to convince the developmental psychologists there to do exactly that, but of course, they wanted to teach about research.

Whenever there is a roomful of people talking reasonably about education, there are many reasonable suggestions. The problem is that soon enough, well-meaning people would wind up designing a system that looks a lot like the one we already have in place.

No one ever agrees to eliminate history, and all agree that mathematics must be useful, even if it has never been useful to them. This goes on and on until the hypothetical system being devised by intelligent people is as awful as the one we have now.

At some point people, and by this I mean school boards, governments, universities, and average citizens, have to get over the idea that there should be any requirements at all in school.

Now I realize that this is a radical idea. Do I mean students would not be required to learn to read or write or do basic arithmetic? No. I mean that after these skills have been mastered, students should be let alone, or rather enticed, to find an interesting path for themselves. The schools ought to be constantly and diligently teaching students to think clearly and should not be trying to tell them what to think about.

We will never change education as long as we hold on to our favorite subjects and insist that they be taught. Everyone has a favorite subject, or has an axe to grind, or has a stake in something not being eliminated. Soon enough it is all sacred, and school is deadly boring and irrelevant.

Anyone who has ever been part of a curriculum committee in a university knows what I am talking about. Everyone fights for their own subjects.

NO to subjects and NO to requirements. Let students learn to do what they want to learn to do. Schooling should be about helping students find a path and succeed at what they have chosen to do.

The politics of history

I was watching C-SPAN yesterday and there was Lamar Alexander, former Secretary of Education and now a U.S. Senator, speaking in the Senate on restoring teaching history to its "rightful place" and making sure that history was part of the NCLB Act. Since he says this sort of stuff all the time, here is a quote from him from 2006, taken from what was pretty much the same speech:

> Just one example of how far we are from helping our children learn what they need to know. The fourth-grade national report card test asked students to identify the following passage, "We hold these truths to be self-evident that all men are created equal, that they are endowed by their Creator with certain unalienable rights, that among these are Life, Liberty, and the Pursuit of Happiness." Students were given four choices: Constitution, Mayflower Compact, Declaration of Independence, or Articles of Confederation. Less than half the students answered correctly that the passage came from the Declaration of Independence. Another question said, "Imagine that you landed in Philadelphia in the summer of 1776. Describe an important event that is happening there." Nearly half the students couldn't answer the question correctly that the Declaration of Independence was being signed.

When it comes to education, politicians never seem to get it. What histories do students "need to know"? None, actually, unless they plan on being historians (or maybe senators). I realize this is a radical point of view, but just like math and science, history is not something anyone needs to know.

Why not? Because knowing what happened in Philadelphia in 1776 does not in fact make you a better citizen, no matter what Alexander says. A good citizen would be one who carefully considered the issues when voting. That would mean being able to think critically and ask hard questions of politicians.

In 1776 we had a bunch of politicians who, if the present is any example, were surely voting for their own special interests. The fact that we, as a country, feel the need to make them into folk heroes does not make it one bit more likely that they were any better or worse than the current politicians who govern us. What Alexander is really arguing for is more indoctrination – informing students what to think instead of teaching them how to think.

Students don't need to know any official facts. They need to know how to live their lives intelligently. It is quite obvious that schools and school reform movements do not have this as an item on their agenda. Rather, just more cramming for tests.

The peril of passion

I recently overheard a mother of college-age kids talking about helping her children find their passion. Then I received an email from a professor I know, about a test someone she knew was creating to help kids find their passion. Numerous books exist on "helping you find your passion," complete with exercises to help you remember what you loved to do as a child, or teaching you how to brainstorm. And then of course there are plenty of academics who write about passion-based curricula.

Sounds nice. Everyone should do what they love.

On the other hand, we have the "exposure" people. Defenders of the existing school curriculum often use the word "exposure" to defend the fact that everyone must take algebra or chemistry in high school or write papers on Dickens in college. How will you know if you are interested in these things if you are never exposed to them?

This puts modern-day parents in a bind. They are torn between racing around to various after-school classes and summer programs and extra lessons that will help their children find their passion and, at the same time, reinforcing decisions made by educators a century ago that their child must study, and do well in school, and learn to love whatever it is they are being exposed to this year.

I feel sorry for today's parent. So much obsession with something they cannot control. They can't fix the curriculum (only Bill Gates can do that – hence Common Core – and it's hardly a fix) and they can't figure out how to help their child find their passion. (Chess club, skating lessons, flute playing, soccer practice – there is so little time.)

So let me make a few observations. I realize I am long past the age of raising children, and that the modern generation of parents takes it all so seriously (while mine just said "go out and play"). But really, let's think.

- Your kid's passion may suck. My son was passionate about being a rock star; I said no.
- My daughter wanted to be a ballerina; I said no.
- My son wanted to be quarterback of the New York Giants (so did I); I said no this time because "really?" it wasn't going to happen.
- My daughter wanted to be a full-time writer; I said she had to learn a real profession, something that would help her eat.

Yes, I know, I am out of touch. I actually vetoed my children's choices of majors in college (English and history). I would have vetoed my son-in-law's choice too (Russian literature) but I didn't get a vote.

We need to realize that high school and college are so out of touch with the real world that the choices they offer (what they expose kids to) is for the most part useless (unless of course they wish to be professors or researchers).

The other options, the passions that we hope our kids will develop, are typically taken from a set of hobbies and are not about realistic opportunities in the real world. They should be passionate about getting a job someday.

In 1962, I chose computer science as my field of study. I didn't know a thing about it, except it seemed to be something new that might matter, and being good at math (which I was) was supposed to be helpful (it wasn't). Nevertheless, I was able to find my passion within that field.

I have found over the years that things that make me angry give me a passion to fix them. First I was angry that computers were so dumb, so I decided to try and fix them. I was angry that they didn't understand English, so I worked on fixing that. And I was angry that they didn't learn, so I worked on fixing that.

Now I am angry that people don't learn anything of value, so I work on that.

What is it that people don't seem to be able to learn? For one thing, they aren't able to make good parenting choices. (If parenting was or is your passion, good luck with courses on that.) Here are some simple maxims for parents:

1. Stop helping your kids find their passion and listen to what they talk about. My son talked about subways all the time, so I helped him work on learning how to do that for a living. He has done quite well at that. You could look him up.
2. Start helping the school system change. Do this by letting your kid learn anything that seems like fun while asking constantly, 'How are you going to make a living from that?'
3. Don't insist on college. Tell your kid to go to work for a few years and decide on college when they know what they want to do. After working in a real-world job they might learn what makes them angry.
4. Let your kids do things on their own. I sent my son out on the subway, or anywhere else he wanted to go, when he was ten. I played on the streets when I was eight. Today's parents would have me (or my parents) arrested for doing that. Good luck with your totally dependent children who need you to help them find their passion and who need you to expose them to things.

In short: don't *expose* and don't *look for passions*; just listen and make good suggestions.

National standards are absurd

I saw this statement from West Virginia Governor Bob Wise, who is the head of the Alliance for Excellent Education (he is a nice guy who I like, but very much in favor of national standards):

"Zip codes are no way to educate America's future workforce."

I found this statement so odd that I wrote to him. He responded:

"Wouldn't you agree that current standards in far too many states are too low to prepare students to succeed after high school?"

What a weird take on the problem. The standards are absurd and students know that. States differ on how effectively they force kids to attend schools that they hate. And, while we are at it, zip codes are indeed a way to manage education.

How do we find out what there is to be in life? School should tell us but it does not. I have come to realize that this is a serious issue in our society. We teach people literature and mathematics and then throw them out into the world, figuring they will know what to do when they get there. In reality, knowledge of literature and mathematics is almost certainly not going to help them. We also fail to ask what we want of our students.

I realized this in a deep way one day when I went to the Santa Fe Indian School in New Mexico. I was trying to get the legislature to give me money to build an online school, which is part of my larger effort to build many new kinds of curricula for high school students.

Going to the Indian School was the kind of politicking one does when one wants a bill to be passed. But once there, I had a realization. Telling these people that we could build a technology-oriented curriculum was not going to be all that exciting for them. I pictured myself as an Indian in Santa Fe and imagined that I wouldn't want my kid on going off to MIT, never to return.

Of course, that is exactly what happens in the segment of society I live in. My children were sent to college and were not expected to return. I am not sure where they would have returned to since I was always moving around myself. But, in hindsight, I am not thrilled that my kids do not live near me and, I imagine, if I were an Indian, I would be very concerned that they stay around so that my culture would not die.

So I asked them questions about curricula that were designed to make them think about what their kids could learn to help their culture survive.

Their answer was *Casino Management*. This surprised me, and then, in retrospect, didn't surprise me at all. Of course that is what they need their children to learn to be good at doing.

We never got to build that curriculum in New Mexico, courtesy of Governor Richardson, who had lied to me about his forthcoming approval of the bill. But it did make me understand something about what is wrong with the national standards movement (apart from its canonization of the 1892 curriculum).

People really are different in different places, and have different educational needs. In Wichita they have an airplane manufacturing industry and no one to teach students how to work in it. In parts of the country there are hotels in the middle of nowhere that can't find anyone nearby who might know how to manage one.

The country is trying to make education one-size-fits-all. Instead, it needs to be local.

Does anyone care that kids all over the world think school is useless?

People get to my column in many ways, but quite often they just type something into Google. Here is a verbatim list of things people typed into Google this week that got them to various columns I have written about school. The country the writers came from is listed first:

US: how many kids hate high school
Romania: hating high school
Morocco: advice to students who hate school
Serbia: hating high school
Guam: school subjects are useless
France: why do students not like school
South Africa: school taught me of how useless aim
Romania: why school is pointless
US: reasons why high school is useless
Kuwait: why I hate high school educational
US: highschool is useless
US: i hate high school and i dont want to be there
Canada: why high school is useless
UK: why do we have education
Canada: why high school is useless
US: useless school subjects
US: school is bad for children
US: useless information learned in highschool
UK: what to do when you hate high school
Spain: hate high school
US: i despise high school
US: what does high school teach you
Canada: i hate high school
Canada: books about kids hating high school
Australia: high school for kids who hate school
Brazil: a great deal of what students learn in school these days is a waste of time
US: high school dont teach you anything
India: academic knowledge taught in high school is worthless if they don't prepare us for own decision

I just thought I'd take note of this phenomenon. While people debate Common Core, or worry about evaluating teachers, or raising test scores, or getting their kids into a good school, try to remember this: most kids are miserable in school. We need to stop teaching the silly stuff we teach in high school and stop creating high schools that make students miserable. Apparently we do this in almost every country.

It is easy to change all this. (Amazingly, the answer is not MOOCs.)

We can change it by letting kids learn what they want to learn in a way that is fun. It is that simple. Technology can help with this. A reasonable curriculum could help with this. Thinking kids have a right enjoy their lives could help with this. Unfortunately, politicians who talk about reform aren't helping, nor are they trying to address the real issues. They never really worry about the kids at all.

The people who typed these things into Google are right. School is useless. It's time to do something about it.

Chapter 3: Teaching the Wrong Way

Cavemen didn't have classrooms

We have begun, as a society, to question what we eat. One line of argument involves cavemen. The reasoning is that modern man evolved over a million years, but agriculture has only been around for the last 10,000 years. The population size grew tremendously as a result of the fact that things like rice, wheat, potatoes, and such could be grown in sufficient quantity to feed large numbers of people. But the consequence of this was a less healthy population. We evolved to eat fruits and nuts, fish, and sometimes some meat. Eating corn chips was never part of the plan. Maybe 50,000 years from now our descendants will need ice cream and assorted chemical preservatives in order to thrive. But today, we are still those people who evolved from the natural selection process of functioning well on what was available to hunt and gather. We are still cavemen, biologically. We just wear better clothes.

I take this idea, seriously and suspect that we are also cavemen mentally. Just as we were evolved to live off the land without excessive alteration to what we find there, so have we evolved to think and learn in a certain way, a way that may not be consonant with how we think and learn in the modern world.

It seems a silly question to ask, for example, but do you think cavemen had classrooms? Did cavemen give and listen to lectures? Did cavemen read and write? These questions matter a great deal, and the fact that we have not asked them has had bad results in the same way that eating corn chips has had bad results.

Why does it matter that cavemen did not have classrooms? The existence of classrooms is based an assumption, that is never discussed but always assumed, that people can learn by sitting quietly and listening. We imagine that people learn by being told the truth by experts and practicing to take tests to see if their minds can retain that truth.

What about the mind of the caveman? Is it reasonable to assume that a caveman was in the habit of sitting quietly and listening to someone who was trying to teach him to know something? The image of that seems so funny that one doesn't even have to add the image of the caveman taking the multiple choice tests that would naturally follow.

Why do these images seem absurd? Because we imagine that cavemen taught their children by example. We imagine that they took them along on the hunt when they were ready and that they practiced, by playing, prior to that. We assume they learned to build shelters by doing simple tasks first, and that they learned to defend against predators by watching and later helping. We don't really have to stretch our imaginations for this, as there are primitive societies where this still takes place today. In fact, prior to the idea of mass compulsory education, like that of mass feeding, we knew how to educate children properly, that is, in the way that their minds were set up to work after one million years of evolution. Instruction in caveman society, indeed in all societies until very recently, was by long-term apprenticeships. Knowing was not valued. Doing was seriously valued.

To put this another way, the caveman's mind was never prepared for, or concerned with, knowing. There was no test. There were no game shows. There was no Nobel Prize. There was action. The winner was the person who brought down the elk or the buffalo. He didn't have to know how to do it, at least not consciously. He had to be able to do it. What knowledge he had was unconscious. He may not have been able to say what he knew that helped him throw a rock straight. He could just do it. He practiced a lot.

On the other hand, we can assume, again from looking at current primitive societies, that he could talk about what he had done. After the hunt, the story of the hunt would be told, perhaps many times. And, we can assume that the story would be told interactively, with many contributing and with remarks from the audience. There probably was no lecture. There certainly was no test. Least of all was there PowerPoint.

Why does this matter? The caveman could not listen to a lecture. His mind was not ready to absorb information in that way. He needed images painted in his mind because his memory was visually oriented. He could easily remember the places he had been and how to get to them. He could remember faces and smells. He could sense danger. Before he had words, he had these things. Modern man has words, but our minds have not changed so much that images, smells, sounds, and surfaces would not be more important as items in memory. Our memories were set up to recall in this way and to use images and senses, long before we knew any words.

And when finally there were words, there were stories. Man has had language (without reading and writing) for hundreds of thousands of years. Hundreds of thousands of years of telling and hearing stories has set us up to be able to tell and hear stories and relate them to experience. We like it. We are good at it. We crave stories.

Sometimes, technological innovations help us to explore who we are. Movies, for example, would work well for cavemen. They liked stories and they liked images. As long as the movies they saw corresponded to their experiences, they would, I am guessing, be able to enjoy them. Technology can enhance the abilities of our basic caveman selves, if it is in concordance with those selves. If it is discordant with those selves, it can be worse than useless.

Modern man needs stories, but he doesn't always get them. Or, if he gets them, they often provide more information than he can possibly absorb. Modern man is not equipped to listen to anyone talk for an hour without interruption, let alone remember much of what that person said. So when airlines blather on about safety precautions, we can safely assume that people cannot remember any of it, because their minds are not set up to absorb information in that way.

What about books? People read books and remember what they read, don't they? Well... no.

People remember some of what they read. They certainly don't remember all the words. If they think about what they read they can remember some of the ideas, or the gist of story. But much is forgotten. Why?

Reading was not a caveman activity. In other words, we do not have hundreds of thousands of years of mental evolution in support of the ability to gain information from books. This brings up the question of what kind of mental activity our evolution does support.

Cavemen needed to know the roles they played in their society. They needed to know how to perform the actions associated with those roles. They needed to have the concept of a goal and they needed to be able to figure out a course of action that might achieve that goal.

Of course, in reality, there probably was not all that much 'figuring out' going on. Most roles and associated actions, and the goals and plans that would help achieve them, would have been developed over long periods of time by their ancestors. The issue would have been the transmission of this information. People learn by copying others and doing things themselves, perhaps with just-in-time advice offered by a more senior practitioner.

The mental apparatus used by the caveman, and thus inherited by the modern man, would have to be all about roles, tasks, goals, and plans, and learning by watching and doing. Just-in-time storytelling would also have a big role to play. Modern man is equipped to do this kind of mental functioning in the same way as he is equipped to absorb the nutrients from fruits and nuts. And similarly, he is not equipped to absorb tightly packed, time-consuming information about what he might need to know in the future should a particular situation ever come up any more than he is equipped to live well from the nutrients in a mocha cappuccino.

Cavemen didn't calculate the surface area of a potato chip

But cavemen did know that F=MA. Cavemen had to know a great deal of physics in order to effectively use a spear or any other weapon. One can assume that they didn't know the language of mathematics (or physics). But mathematics is just that, a language. And it is language that gets in the way of our understanding of how the mind works. The mind understands well enough that more force makes more impact. Force and impact are very basic notions – they are some of the indices that the mind would have to use in order to function. Formulas are for those who want to discuss and analyze how the mind functions, not for its actual functioning.

The caveman was probably not conscious. There is ample evidence to demonstrate that consciousness is a fairly recent human phenomenon. If we teach to the conscious, if we say how to do something, or worse, teach the theory of how something works, rather than show how to do something, we lose the student, because his mind does not work that way. If experience is separated from knowledge, and if what we teach is not at all about doing, then we are teaching to the conscious. Conscious people may make good intellectuals, but those intellectuals are unlikely to become practitioners.

Call me crazy, but I think we have plenty of intellectuals. Teaching people to work together, reason about new situations, and achieve their goals, just as cavemen did, is what education should be about.

The library metaphor

Certain metaphors dominate our society and our way of looking at the world. George Lakoff, a cognitive linguist, has pointed out that using language such as, thinking is like "giving birth," or seeing competition as "war," cause us to actually believe that these ways of talking about the world are the right ways to understand them. Competition really becomes war. We need to "kill" our opponent in the match. Sometimes these metaphors are helpful. Seeing time as money may help us "budget" our time better. But sometimes, they are disastrous.

One such disastrous metaphor has dominated thinking about learning for a very long time. We need to get over it if we wish to ever see schooling become relevant in the "knowledge society." I am talking about the metaphor of knowledge as akin to something to be found in a library.

Libraries have been around for a long time. For generations, knowledge was contained in libraries, or so it seemed. But, in fact, this was never true. It didn't matter much, until recently.

Concomitant with the idea that knowledge is contained in libraries is the idea that knowledge is found through search. In the old days, when people actually went to libraries, there were card catalogues, which were created with arcane notions such as the Dewey Decimal System that helped searchers find books that had been properly catalogued. But we don't need that stuff anymore, because we have Google. Search has gotten easier, but real knowledge hasn't changed.

The problem is that both the library metaphor and the search metaphor have misled us in serious ways. The consequences of that will take a moment to explain.

When everyone agreed that libraries contained all that mankind knew, educational systems evolved in such a way that mastery of what other people had written passed for education and hence erudition. Thus we have the Great Books, and the original conception of universities as places to read what great thinkers had written. The concept of testing to see if one had learned what these greats had written follows from this. Given that information retained from books can be measured, as can the sheer number of books read, school became a kind of competition to see who had retained the most. The winners go to Harvard.

Behind all this is the idea of the mind as a kind of library. Libraries are where knowledge is stored, so the mind must be a particular kind of library and education must be about filling that library.

In reality, the mind is no kind of library at all. We lose "books" we have filed away, we mush together similar "books," and worst of all, we really don't consider it our job to know what we know. The job of the mind is to deal with what is going on at the moment. The mind is goal-driven, not knowledge driven. Knowledge is useful to the extent that it helps us accomplish goals. In fact, any child knows this. But the school system does not. So when it finds a body of knowledge it likes (like algebra), it requires that those books get stuffed in the library and checked out from time to time. The students ask: Why do I need this? When will I use it? What goal will it help me accomplish? Since the actual answers to those questions are "you don't," "never," and "none," the system refuses to answer the question and instead says things it can in no way prove, like "it will train your mind." We are stuck in a bad metaphor. One that thinks knowing the works of Dickens is what knowledge is, when in reality, knowing what to do in a given situation is what knowledge is. Procedures matter. The more processes you know (that is, the more you can execute), the more you can do.

School is not about doing, despite scholars from Plato to Dewey saying it should be, because of the library metaphor. Doing is hard. It is hard to do and it is hard to teach. But if knowledge is about storage, then school becomes easier to manage. If the best students are those who store and search well, then we can figure out who goes to Harvard. But if knowledge is the service of the achievement of daily human goals, then knowledge might be something hard to explicitly state and to measure.

We must get rid of the library metaphor, or school will always be the same: an experience to be endured rather than relished.

How to build a culture of illiteracy

I was visiting my five year-old grandson this week and he showed me the two books he had been assigned to read by his kindergarten teacher. Milo reads fairly well for a five year-old, but this has nothing to do with his two months of kindergarten. It surprised me to see that they are teaching kindergarten kids to read. I guess the obsession with test scores has made New York City push the kids harder and faster. I would rather see him do other things in school, but there is no harm in teaching him to read. Or so I thought, until I had him read these two books to me.

The first was about a nonsensical creature called a jiggeridoo. All the sentences were of the sort "jiggeridoos like to play." Each page had a picture and a sentence like that. I was bored and so was he. What happened to the idea that reading should be fun and even remotely educational? What was he learning from reading this nonsense?

The second book made me long for the first one.

The second was about a character named Eddy, who liked to eat things based on their shape, apparently. So Milo was busy sounding out words like square or triangle because pizza is triangular in shape. This book was both boring and annoying. The author was obviously trying to teach shapes, which is a dull and rather unimportant task, and was doing so through a ridiculous book.

From this Milo is learning that books are boring and tedious. He will survive this because his parents are literate and will be his real teachers. But I started worrying about kids who have parents who don't encourage reading at home. The message the school is sending is that reading is a useless experience teaching nothing worth knowing. No wonder illiteracy is such an issue these days.

Perhaps we should stop worrying about test scores and start worrying about whether kids like to read. Just a thought.

The history teacher or the football coach?

I play softball in an old guy's softball league in Florida. I started playing a few years ago and I discovered I wasn't really very good. This was a bit surprising, since I had played in university softball leagues while I was a professor, and had only stopped playing in my forties. I wasn't a bad player then and there hadn't been that long a hiatus. And now I was playing against people a good deal older than myself, since I am rather young as recent Florida transplants go. I used to be a good hitter and now I wasn't. The reason was easy enough to understand. In the university leagues they play fast pitch. A batter has a second or so to decide about swinging. It is all instinct, at least it was after having played for forty some odd years.

But in Florida, old guys play slow pitch. The pitcher throws the ball in a high looping arc and it is a strike if it lands on the plate. Quite a different experience from trying to hit a ball that is zinging by your head. Should be easier, no? Not for me. It took a bit of thinking to figure out why.

I analyzed how I was swinging, when I was swinging, what kinds of pitches I was swinging at, and I came to many different conclusions. I realized I needed to wait longer before I swung. I realized I had to stop swinging at inside pitches (the ones that almost hit you). I realized that I had to stop swinging at pitches that looked good but yet dropped in front of my feet. I realized I had to see the ball hit the "sweet spot" on the bat. I realized, in fact, I needed to change my whole approach to hitting.

Okay. I realized a lot. I had come to many conclusions. Now what? Just do it, right? Aha. Not so simple.

You can't just do what you know you should do. Why not? Because your unconscious isn't listening to what you have to say.

You can tell yourself to do this, that, and the other, but your "self" isn't listening. Did you ever wonder why what you learned in school isn't still in your head, or why you can't remember what your wife wanted you to get on your way home? Or, why the things you hear about that will help you improve your business or make more money or be a better person don't actually ever get executed? The answer is simple: you can't learn by listening – not from teachers, not from your wife, not from helpful suggestions from wise people, and not even from yourself.

Why not? Because it is your unconscious that is in charge of executing daily activities – from swinging a bat to driving home to talking to people you want to make an impression on, to getting along with your wife. Your conscious can make decisions, but your unconscious pretty well does what it is in the habit of doing. The unconscious is a habit-driven processor. It says stuff you didn't mean to say, comes to conclusions you didn't know you believed, and in general, is running the show.

Bad habits, as they say, are hard to break. Actually, all habits, good or bad, are hard to break. A new swing is really hard to develop, as is a new way of selling, or a new way of treating people, or driving a new route home. Education that tries to instill new habits, where there are no old ones, tends to work rather well. Young children learn from their parents by unconsciously copying everything their parents do, including things the parents would just as soon not see their children doing.

The real value of education is in the creation of new habits. This can only be done in one way. The unconscious only learns in one way. It learns by repeated practice. The only teaching that really works is the observation of good role models and the kind of mentoring that helps someone execute better while they are trying to copy what they see others do.

And this brings us to a key question about education. How is a high-school football coach different from a high-school history teacher?

Before we attempt to answer this question we need to consider why it is an important question to consider. In general, I think most people would agree that the behavior of these two types of teachers is likely to be quite different. In our mind's eye, we see images of yelling and crude behavior versus refined lecture and discussion. But, let's get beyond the superficial stereotypes and think about what they teach rather than their style of teaching.

The history teacher at his worst teaches facts, and at his best teaches careful analysis of sources of facts and consequences of events.

The football coach at his worst teaches that someone could never possibly do something, because they are fundamentally bad at it, and at his best coaches someone to do something better than he or she could ever do before.

The history teacher teaches the conscious. The football coach teaches the unconscious.

Education – or more accurately, school – is a conscious affair. We discuss history – we don't do history. We are frustrated when what we learn in school is forgotten years (or even weeks) later, but we fail to understand why it is forgotten. We see ourselves as deserving of blame when this is far from the case. All those facts, gone. Algebra problems you were once good at now look like they are written in Chinese. What happened?

The problem lies in the nature of school itself – what is taught and how it is taught.

If you really want to learn something, or to teach something, it is important to understand what happened. It is also important to understand why the football coach is more successful. Thirty years later his charges can still catch a pass, and they have not forgotten how to tackle. They remember it all.

What is the difference? And, more importantly, what can we learn about learning by examining that difference?

One place to start is in understanding the difference between conscious and unconscious knowledge.

The football coach doesn't need players who can discuss football, he needs players who can execute. Because he is interested in execution, he does not dwell on the issue of passing an exam about football or being able to write an essay on football. In fact, many people could pass such an exam or write such an essay, who could not begin to be able to catch a pass. (Often they are called sportswriters – or fans.)

We can talk about what we know consciously. This ability to discuss conscious knowledge does not relate one way or the other to the ability to execute using unconscious knowledge. Conscious knowledge has next to nothing to do with unconscious knowledge.

How do the conscious and the unconscious interact? As long as we see ourselves as rational beings who can think logically and make carefully reasoned decisions about our daily lives, then education indeed should be about the promotion of reasoned deliberation and the gaining of knowledge that will enhance our ability to reason. But, suppose this conception we have of ourselves and our ability to reason logically is simply wrong? What if it is the case that we can't actually reason logically at all?

All of our education system depends on the "pure reason" assumption. The assumption is that we teach conscious minds that are capable of logical thought, can remember what they were told, and can think clearly about the consequences of their actions based upon all this knowledge. But in reality it is the unconscious mind that can be taught. It is the football player who really learns. The history student simply learns for a short time, and then forgets what he or she has learned, and is incapable of ever making serious use of what he or she has learned. He can sometimes recall it while playing Trivial Pursuit or Jeopardy, but the usefulness of the knowledge almost never extends to helping with decision-making.

The conscious isn't really capable of retaining what it has learned in any useful way, in part because what it has learned usually does not relate to anything it will later be called upon to do. The unconscious, on the other hand, is always ready to learn in the service of doing, because that is what it has always done.

Can you tell if it is school or prison?

I read a blog by a parent concerned with his daughter's education that pointed out that schools and prisons look alike, but there is much more in common between school and prisons than their looks:[1]

1. Students/prisoners (s/p) must stay in the place they have been assigned unless given specific permission by the guards/teachers (g/t).
2. S/p may eat only with permission from g/t.
3. S/p may go to the bathroom only with permission from g/t.
4. Assigned tasks must be completed by s/p.
5. Questioning the task you have been assigned is not allowed.
6. Expressing a point of view contrary to the g/t about rules is not allowed.
7. The g/t may humiliate an s/p at any time.
8. The s/p may intimidate and terrorize other s/p.
9. All recreation is supervised by g/t at specified times.
10. Reading material is deemed suitable or not suitable by g/t.
11. All visitors must be vetted prior to visitation.
12. Failure to follow the rules will result in punishment.
13. Failure to behave properly may add extra time onto one's sentence.
14. Approval by g/t is determined by extremely arbitrary standards.
15. Freedom of expression is strictly controlled.
16. Dress codes are strictly enforced.
17. Getting the g/t to like you will make your time go more easily.
18. Resting is not allowed.
19. Pursuing one's own interests is not allowed
20. Deciding you have better things to do is definitely not allowed.

I realize that not all prisons and schools are exactly the same in all this, but you get the idea.

---

1 Essential Emmes http://essentialemmes.blogspot.com/2010/02/genius-on-education-roger-schank.html

Why can't we express emotion properly? A call for new school standards.

I was in college when John F. Kennedy was killed. We were finishing lunch at the fraternity house. We walked around in shock, or were glued to the TV. Those were simpler times. Less political. Everyone loved the President.

Not knowing what to do with myself, I went to my next class. All of the class was there. It was an economics class. The professor thought it would be a good thing to discuss the potential economic impact of Kennedy's death.

I never took that class (or economics as a subject) seriously again. I didn't know what I thought he should have done, but I was pretty sure it wasn't a discussion of the economics of assassination.

I was reminded of my feelings about those events and that class by the recent Boston Marathon Bombing. News coverage is more elaborate now, but the newscasters said this time, as they did then, most of the time, that they had no idea what was really going on.

But there was something different this time. While everyone I knew then was simply in shock or angry or numb, the people in Boston, at least according to the TV coverage, were singing, waving flags, applauding, and going to Red Sox games where Red Sox songs were being sung. There was a lot of cheering for the good old U.S.A., and lots of Boston pride.

Didn't people die? Weren't people horribly injured? I would have expected more crying and less cheering and singing.

It is possible that I am out of touch after all. The world changes as you get older, as does young people's behavior. But this is a column about education, and I can't help but see this as another failure of our education system.

Why would it be wrong for children to be discussing their feelings and thinking hard about what can be done to prevent such horrible events? Or thinking about why people do things like this?

Why can't adults think clearly about these events? *Heightened security at the London Marathon?* Did someone expect a series of marathon attacks? *Heightened security at airports?* Maybe in Boston if the bad guys were leaving the city, but at every U.S. airport?

This is not meant as a criticism of Boston. Look at this headline from *Yahoo Sports*:

> Citi Field breaks into 'U-S-A!' chants after Boston Marathon bombing suspect is taken into custody

Of course I am not the only one to be appalled by this. From *21st Century Wire*:

> How did Friday become such a huge 'patriotic moment' for the people of Boston? Was this some kind of victory for America?

My answer to all this is simple enough. *It's school.* In school, where we should be discussing things, expressing points of view and trying to figure things out, we are instead preparing for tests. We are learning "right answers" and one of those is that the U.S.A. is the greatest country on earth. We are not learning how to think. We are not learning how to express emotions in a reasonable way.

This is, of course, not limited to the U.S. I received a letter from Spain yesterday from a mother concerned about her son. She said (among other things about how her son hated school):

> As most kids his age, he loves music and sports (I do encourage it as far as I can). He also writes beautifully, I know, because he complains in writing and it always impresses me how successfully he does it. But at school they don't encourage it at all as they're always more concerned with spelling and so on than with the content. So he just complains in writing instead of using that talent more creatively.
>
> At school his results in music are always low as they value the theoretic part of exams (you'd never believe what all that is about), so, the practical part of the subject is always buried and he loses interest in that also. Same goes for sports. I wonder if our Spanish sports talents such as Rafael Nadal were successful in theoretical sports at school?

Of course, if this boy was engaged in music and writing it would be because he had been emotionally engaged. Learning is emotional because we care about what we learn, and get excited about what we learn, and share what we learn with others.

Well, that would be the case if school were about what people wanted to learn. Yesterday there was some discussion in the press about the college readiness of high school students (as there usually is, this being the topic of the day), and a report from the ACT (the testing people, for my non-U.S. readers) complained about how students aren't college-ready. The translation of this is that they need more test preparation (sold by the ACT, of course). Here is a paragraph from that report:

> Especially at the high school level, where there are differing degrees of familiarity with the improved standards, state and local efforts to implement the standards have not yet achieved their goals. This suggests that not enough teachers are yet ready for the necessary changes in curriculum that are likely to accompany

> the switch into a classroom environment driven by college- and career-ready standards.

The translation of this paragraph is that they want more testing, more test preparation, and to make sure the new standards in math, science, etc. are being met.

I would like to call for some new standards too.

I would like to see a happiness standard. If kids aren't happy in school, the school is failing and we need to fix it.

I would like to see an emotional readiness standard. If kids can't express what they are feeling – in writing, in discussion groups, to friends – then they need to learn how to do so. If we express emotions about bad guys getting killed by dancing and waving flags and singing, we have clearly missed the lesson on how to express empathy, relief, and fear.

I would like to see a clear thinking standard. We need to teach people how to react to events they don't like by planning new courses of action that make sense. Learning to plan a course of action is very important, but if that happens in school, I missed it.

And, lastly, I would like to see the following standard:

Schools are not allowed to bore their students so badly that they see school as being irrelevant to real life.

Here is a question that comes from the ACT online practice tests. I fell asleep reading it. But I learned that no one can express or feel real emotions because any real emotion you might have while taking this test would not be dealt with well by the school administering it. It starts like this (and goes on and on):

> **Passage I**
> Unmanned spacecraft taking images of Jupiter's moon Europa have found its surface to be very smooth with few meteorite craters. Europa's surface ice shows evidence of being continually resmoothed and reshaped. Cracks, dark bands, and pressure ridges (created when water or slush is squeezed up between 2 slabs of ice) are commonly seen in images of the surface. Two scientists express their views as to whether the presence of a deep ocean beneath the surface is responsible for Europa's surface features.
>
> **Scientist 1**
> A deep ocean of liquid water exists on Europa. Jupiter's gravitational field produces tides within Europa that can cause heating of the subsurface to a point where liquid water can exist. The numerous cracks and dark bands in the surface ice closely resemble the appearance of thawing ice covering the polar oceans on Earth. Only a substantial amount of circulating liquid water can crack and rotate such large slabs of ice. The few meteorite craters that exist are shallow and have

> been smoothed by liquid water that oozed up into the crater from the subsurface and then quickly froze.

There is additional material from "Scientist 1" and then another paragraph from "Scientist 2."

It is followed by exciting questions such as:

1. According to the information provided, which of the following descriptions of Europa would be accepted by both scientists?
    A. Europa has a larger diameter than does Jupiter.
    B. Europa has a surface made of rocky material.
    C. Europa has a surface temperature of 20°C.
    D. Europa is completely covered by a layer of ice.
2. With which of the following statements about the conditions on Europa or the evolution of Europa's surface would both Scientist 1 and Scientist 2 most likely agree? The surface of Europa:
    A. Is being shaped by the movement of ice.
    B. Is covered with millions of meteorite craters.
    C. Is the same temperature as the surface of the Arctic Ocean on Earth.
    D. Has remained unchanged for millions of years.
3. Which of the following statements about meteorite craters on Europa would be most consistent with both scientists' views?
    A. No meteorites have struck Europa for millions of years.
    B. Meteorite craters, once formed, are then smoothed or removed by Europa's surface processes.
    C. Meteorite craters, once formed on Europa, remain unchanged for billions of years.
    D. Meteorites frequently strike Europa's surface but do not leave any crater.

Now ask yourself: can people who have gone through a system that is this devoid of emotion, that fails my standards so badly, and that is so irrelevant to anything they will actually do in life, ever know how to express emotion or find solutions when bad things happen.

Teachers' despair: we cannot afford to be focused on training intellectuals

As part of a presentation to teachers in Mexico City, Telefonica of Spain set up a forum for teachers to ask me questions in advance.[2]

The forum is in Spanish and my Spanish is minimal at best, so they sent me translations of the questions and comments that were posted. What I am struck by, as always, is the difficult situation in which teachers find themselves these days. It really doesn't matter what country a teacher is in, they are always faced with two truths:

1. They are not quite sure whether what they are doing in school is really the right thing to do.
2. They know they have very little power to change things.

Here is one question I got, for example (excuse the awkward translation, I received them all in that form):

> We learn something new every day, depending on our attitude towards learning, and even if we are not going to put it into practice, we need to take it in as part of our general knowledge. For example, why is philosophy important for someone who is going to study engineering? There is some material that is simply useful in life. Is this assessment correct?

I can't help but feel this teacher's pain when reading this. I am saying, as usual, that we only learn by doing, and this teacher is trying to figure out how what he or she is doing is still ok. "If we don't put it into practice, isn't it still okay to teach?" Now of course the answer for me is "no," since I believe that we only learn by doing, but consider the teacher. The teacher stands up in front of class trying to teach general knowledge that will never be used. The teacher's hope is that philosophy will be of use somehow to someone and that the "general knowledge" that is the staple of the school system will somehow turn out to be useful, even though this teacher isn't really so sure it will.

Consider this next question:

> Learning depends more on the person doing the teaching, on the strategy and methodology applied, than on the student. This is because a good methodology can make the student take interest in what he/she is doing and be enthusiastic. Is that right?

2 The forum can be found here: http://encuentro.educared.org/forum/topics/solo-se-aprende-haciendo?groupUrl=solo-se-aprende-haciendo

Here we have another teacher saying that a good teacher can make students excited about anything, so isn't that a worthwhile thing to be doing? Well of course it is. Turning students on to things they didn't know about and getting them to care about them provides enjoyment, and could possibly have a large affect on the rest of the student's life. What's the problem, then?

The problem is well expressed by this next question:

> It is possible to learn almost anything. All we need is motivation. We must try to somehow involve, motivate and encourage students to participate in their lessons... Is it possible to learn through practice, even when what is learned is of no use to the student?

This teacher is willing to accept the fact that what is taught in school may be completely useless to the student's future life. I, for one, find that idea very difficult to accept. I realize that teachers teach what they are ordered to teach, but what must it be like to teach material that you know is completely useless to the student?

I ask this question as if I don't know what it is like, but of course I know it all too well. The reason I became an education radical is that I was teaching a course in semantics at Stanford and realized within a few days that no student in the class cared about, or would ever make use of, what I was teaching. They were simply required to take semantics. I knew right then that I needed to rethink.

I will now consider the last (of the ones I have chosen to write about) three questions together:

> 1. Would it be wiser to focus more on the theoretical basis than on practice? Students show more interest in classes in which the outcome is an object constructed upon a scientific foundation.
> 2. It is extremely important to find the reason behind what we teach and, often, this raison d'etre is the source of knowledge or of its use in other sciences or fields of knowledge.
> 3. Is it possible to remember what they have heard in a reading if it is truly significant to them? If what is read motivates the reader, does this mean there is a greater chance of learning it? Or do we only learn by doing?

I get an overwhelming sadness from these questions taken as a whole. These teachers are focused on teaching science and basic knowledge and great books. This is what they do and it is what they have always done. They ask if there isn't some use to it all and of course there is. This is how we create intellectuals. Intellectuals worry about science and general knowledge and philosophy and great literature. Intellectuals can discuss these things and enjoy doing so. They may use them or they may not but it is part of the well-rounded education of an intellectual.

My question is about the percentage of intellectuals out there in world. I find it hard to believe that our school systems in every country are geared towards the creation of intellectuals. I am sure 90% of all students have no interest in becoming intellectuals. They would like to learn to earn a living and how to take care of their families and how to be good citizens and how to have good relationships with people. They would like to know how to function in the world. While we can kid ourselves that making them read Don Quixote, or read about the glories of the Spanish Armada will somehow contribute to their greater development, this is just wrong and irrelevant to their lives.

Our education system was designed to create intellectuals. In the U.S., it was designed by the President of Harvard (in 1892). He wasn't interested in the average person. He was interested in the elite who would attend Harvard.

All this must stop. We need to focus on getting the general population to be able to think clearly. This does not mean teaching algebra and chemistry and pretending that such things teach clear thinking. It means having students practice making decisions and understanding the consequences of those decisions. It means having them come to a conclusion about something they care about by learning how to examine evidence. It means having them learn to create a plan that will help them get what they want and then executing that plan. The average person does not need to read Descartes, no matter how much we rationalize to ourselves that Descartes said some things that might be of use to the average person.

I know teachers can't change the system by themselves. But they need to band together and try to make some changes, or another generation will be lost.

The myth of information retention

The other day I heard about a professor who tests his students online continuously during his lectures, and found that they retain more information than students to whom he lectures without the tests.

My first thought was shock that he put students through this (although lecturing is so dull that maybe this makes it more fun), but then I began speculating on the concept of "retention of information."

We all, it seems, wish we could retain more information, and most people chastise themselves for forgetting things. I forgot to get English muffins at the store the other day and I have been chastising myself because there was a practical consequence to that – I can't have them for breakfast.

But I have also forgotten nearly anything that I learned in college. I don't chastise myself for that, since anything I really needed to know I have used a zillion times since, and anything else, well, I didn't need to know it.

However, I did remember about the bubonic plague – you hear that story a lot. Yesterday, news came out of the U.K. that there never was a bubonic plague and the actual plague they had was not caused by rats.[3]

So, I had retained information on the bubonic plague, but it was wrong. Jeopardy, Trivial Pursuit and, most of all, endless testing by schools and the anything-but-student-centric College Board, have convinced a generation of Americans that retention of information is the key to something very important. I am not sure what. Good grades I suppose. And good test scores. It is well to remember that tests, especially those in college, are usually an attempt by professors to ensure that students at least try to pay attention to what they are hearing about. We don't learn much from lecturing and every professor knows it, so retention of information has become an idea that professors force students to dwell upon.

I looked up some tricks for retention of information that you can find on the web. Here are some excepts from one site:[4]

- Focus your attention on the materials you are studying.

3 Black Death might have been airborne, U.K. scientists suggest http://www.washingtonpost.com/blogs/compost/wp/2014/03/31/black-death-might-have-been-airborne-u-k-scientists-suggest/

4 11 Great Ways to Improve Your Memory http://psychology.about.com/od/cognitivepsychology/tp/memory_tips.htm

- Utilize mnemonic devices to remember information.
- Elaborate and rehearse the information you are studying.
- Relate new information to things you already know.
- Visualize concepts to improve memory and recall.
- Teach new concepts to another person.
- Pay extra attention to difficult information.

The gist of these suggestions is that the key to the retention of information is to memorize better. Some of these are perfectly reasonable sounding, but completely useless. If you can't relate new information to something you already know, then in effect you can't even hear it. Learning and listening depend upon retrieving what you have experienced yourself and checking to see how your experience relates to what you are hearing. This is how conversation works and it is why we always have something to say back when people tell us stuff (unless we simply don't care what they are saying). In other words, memory and learning are natural processes. Giving people a list of things they do unconsciously is not very useful.

Had the second to last line been "talk about new concepts" instead of "teach new concepts" it would have been a slightly more useful suggestion. But "teach" is wrong. If we weren't excited enough about what we heard in a lecture to want to talk about it with our friends, then we have no chance of remembering it, much less of teaching it to someone else.

Here is another web excerpt I found on the same subject:

Strategies To Improve Memory and Retention
GULP - GULP is an acronym for an effective four-step process to improve short and long term memory.

Step 1: G - Get It
Step 2: U - Use It (sing or chant it)
Step 3: L - Link It (make an acronym link; alphabetize it)
Step 4: P - Picture It

This one just made me laugh. Alphabetize it? Is school so awful, and lectures so terrible, and studying such a miserable experience, that we must resort to alphabetizing every new thing we hear? Or chanting it? Who writes this stuff? (This one is from a university web site.) The answer is obvious. Teaching is really broken. In fact, teaching is so bad that not only do we not know what to teach (stuff you can't remember), but we also insist that you remember it. No one ever explains why you need to remember any of this, of course. ("For the test" is the obvious answer.)

Here is another one:[5]

> **Memory Retention Rates Tell You How To Learn**
> *Reading Is Not The Only Way To Learn*

And I thought reading was not even one way to learn. I guess I was wrong. I thought practice, thinking, and experience were how we learned.

> *Memory Retention Is Based On Pressure*

It is? Do we push on our head in order to learn? Stress ourselves out in order to learn? Stand up on stage and recite what we have learned? Actually I think this last one is what the author intended. It is a good way to memorize a song or a part in a play, of course. What it is has to do with education eludes me.

And finally one last one:[6]

> **How To Memorize And Recall More Information**
> There is no "best way" to take notes – you need to experiment and test what works best for you. One thing is for sure – when highlighting a book – don't highlight every single line, only highlight what you are sure you're going to have difficulty remembering.

This one was written by the author of a book called *Get the Best Grades with the Least Amount of Effort*. This is a student guide that has been sold to thousands of students in more than 30 countries and translated into four languages.

Good. Now it is clear. No one cares about retaining information with the exception that we care about grades and test scores. We care about grades and test scores because we are forced into taking tests and working for grades by an education system that has abandoned the idea that we learn for any other reason.

Let me make a radical suggestion. We learn so that we can do something we couldn't do before. One of those things ought not be test taking, but in our world that seems to be the only one that matters. How sad.

5 http://www.timelessinformation.com/memory-retention-rates-tell-you-how-to-learn/
6 http://www.howtolearn.com/2011/03/how-to-memorize-and-recall-more-information/

Exposure, cultural literacy and other myths of modern schooling

I received many responses to a column on high school.[7] I will attempt to answer them all by answering just one:

> Dear Mr. Schank,
>
> I found your recent op-ed in the Washington Post ("Why kids hate school – subject by subject") spot-on. Your comments about foreign language instruction, in particular, were quite lucid, and as an ESL teacher (and someone who only learned Spanish by moving to Spain, despite five years of Spanish class in grade school), I can attest to the frequency with which students arrive having studied grammar for years in their home countries without being able to manage a simple conversation, or being able at best to string together a series of formulaic, overly-practiced sounding responses that native speakers rarely use.
>
> And so on with your discussion of the other subjects. I'm curious, though, to learn what you think about students' more general cultural education – their knowledge base about the world. Do you feel that students should come out of the educational system with some sort of fluency in the various subjects? How would something like this be accomplished? It seems like there's something to be said for having familiarity with major historical events, some canonical works of literature, some understanding of how plants work.
>
> Also, how do you see arts education as fitting into this?
>
> Thanks again for publishing and spreading your ideas. Hopefully my questions don't come off as too uninformed – were I somebody with more free time, I'd while away the day looking into your blog and published writings further. Let's call it a long-term project.
>
> Take care,
>
> Kevin Laba

My response

Dear Mr. Laba,

Thank you for your question.

Of course one can make a legitimate argument for the idea that every person should know everything that matters or might matter. Works of literature? Why not? What

7 https://www.washingtonpost.com/blogs/answer-sheet/post/why-kids-hate-school--subject-by-subject/2012/09/06/0bf1acc4-f5d6-11e1-8398-0327ab83ab91_blog.html

harm could Dickens do, really? Everyone should know about World War II. How could one be a citizen of the world and not know about that?

The problem is that once you accept that idea, two things happen and both are bad. The first is that you implicitly accept that telling (or reading) are the means by which students will "know" about these things. But that model doesn't work. We don't remember what we are told for very long. And if we do recall some information, one needs to care about it, use it, do something meaningful with it, in order to have a deep understanding of something, and that just isn't how school works.

School doesn't work that way, in part, because of the second bad thing. Once we think there is important stuff to know, someone is going to make a list of exactly what that is and you get books like "what every second grader must know" which, if I remember correctly, includes Eskimo folk tales because they are "cultural knowledge." The list is long, so in the end someone decides what matters most, and that is how we end up with the curriculum we have.

We don't need to do that any more. It is possible to build thousands of curricula, and because they can be offered anywhere once they are built, students could learn what they are interested in learning. "One size fits all" is a very old idea for education and one that is very convenient for governments, book publishers, and test makers.

I, for one, never wanted to know how plants work. I never cared. But then a couple of years ago, I did care, because of some Artificial Intelligence I work I was doing. So I called a plant biologist I know and asked. I realize that not everyone has the luxury of doing that, but in the age of the web one can pretty much find out what one wants to know.

The real issue is: can you understand the answer? The jobs of schools and teachers should be to cause students to think hard about things they care about. Thinking is thinking. If you learn to think hard about human memory and learning, you can understand a biologist when he speaks clearly.

As for arts education, I have, of course, the same point of view. Those who love it should do it. Those who would like to have a passing knowledge of it should be encouraged to do just that. We can't force people to listen to lectures about paintings or listen to music that doesn't interest them. Well, we can, but it never works.

The key word, the one I have heard again and again in counterarguments to my ideas, is the word "expose." Some very intelligent people have asked the question about how one would know if one wanted to be a chemist without being exposed to high-school chemistry.

I find it an odd question. Prior to the age of 16, a child does a lot of living and has plenty of time to express his or her interests to parents, friends, and teachers (or the web).

Someone who might be interested in chemistry would be asking questions about how the world worked long before being forced to balance chemical equations.

School is the wrong venue for "exposure." In school there is very limited exposure, actually. We expose students to what was intellectually fashionable in 1892. We don't expose them to business, law, medicine, engineering, psychology, and hundreds of other subjects because they didn't teach them at Harvard in 1892.

We need to teach thinking and get away from the idea of "important subjects." There aren't any really.

Reading and Math are very important, but I am not sure why: a message for those who can't see very well

My daughter recently wrote an article about how her vision-impaired daughter was being treated by the New York City education system.[8] She is a smart kid and my daughter wanted her to be in the smart kid class. But, since she can't see all that well, it was difficult for her to pass the required tests to get into this class.

My daughter was, and is, concerned with her daughter's reading ability. In order to pass the tests she needed big print texts. Her daughter reads very well, but that doesn't mean she can see the tests very well.

But this article is not about my daughter, nor about my granddaughter. It is about reading.

The problem for people who can't see very well is that they are expected to read a great deal in school and they are therefore in a very difficult situation. The blind have books in Braille and signs in Braille. This sounds reasonable but it really isn't. What is with this emphasis on reading?

Throughout the history of education, reading has been a big component of the system. Since religions are mostly responsible for how school developed, this is not surprising since religious texts are typically a big part of religion.

When people are concerned with the education of blind children they worry about classrooms that accommodate them, and books that accommodate them, and various teaching materials that can be made to accommodate them.

Now, while my granddaughter does not see very well, I personally have never taken much note of it. Why not? Because I talk to her. She talks to me. She sees well enough to grab my hand when we walk and to notice and deal with things that she encounters. She has lots to say. If this child had been born 5,000 years ago, she would not have been expected to go out on a hunt I suppose, but she would work and function well enough with the other kids in the village and could do whatever she wanted to do. She can talk and negotiate in the world. She is very smart. In the ancient world she would never have been seen as being seriously impaired.

---

8 Twice Exceptional http://www.thebigroundtable.com/stories/twice-exceptional/

But, in our society we have reading and math tests. And then, we have more reading and math tests. We all agree that math is important for reasons that have eluded me and we memorize equations because someone said we have to and then promptly forget them. When I attack math, I have many supporters, but I never win that argument because the tests makers, book publishers, and the people in charge of the system all think math is very important because it just is. ("It teaches you to think" being the usual argument with no evidence provided.)

So, now, I will make an even more ridiculous argument which will be ignored by most of the population. Reading doesn't matter either.

There. I said it. Yes, I know we have set up a system where reading matters a great deal. (Indeed, I am part of that system. I write books. I build online courses that require reading, and so on.) In our world, I type this and someone else reads it. The internet has made reading even more important now than it was when I was a kid. When I was kid it was important to read so we could read *The Scarlett Letter* and *A Tale of Two Cities* and *The Rime of the Ancient Mariner*, all of which I have forgotten, did not find engaging, and have no idea why I was required to read.

Well, actually, I have a very good idea why I was made to read them. The schools were designed by intellectuals to create more intellectuals. Intellectuals discuss literature. Maybe they used to discuss mathematics, but they don't any more. Intellectuals don't discuss science much, so science is on a back burner in schools. Intellectuals do discuss history, so history is taught with great devotion.

(As I write this there is noise being made outside my office. It is being made by people who are working – building things. These people go to school too, but school is not meant for people who want to build things or even for those who want to discuss things. It is built to ensure that the people outside my office feel they are failures because they had bad math and reading scores in school.)

As I have said many times before, we teach the wrong things in the wrong way.

What should school look like? No reading, no math. Yes, I know, that is insane. No one could possibly think this. But, give it a shot for a minute.

If there was no reading and no math what would happen in school? Could we let kids do what interests them? (No, I don't mean video games although anything can serve as an avenue into learning and thinking.) Could we help them find their interests? If reading interests them, by all means teach them to read. The same with math. My grandson, the brother of the granddaughter who can't see very well, is set to go off to a math and science middle school. I asked him if he likes math and science. His answer was that he didn't actually like math and I discerned from what he said about science that he had no idea what science actually was about, thanks to the "science" they teach in school.

When I asked him what interested him, he said "robots." We talked more and I got the idea that he liked learning how things work and engineering would be his choice if anyone actually gave him that choice.

I can hear the chorus now. "But engineering requires reading and math." Well, not necessarily. In order to learn how to be an engineer you do not have to read at all. The way people have learned throughout history is from each other and they learn from each other by talking and by asking and by getting good advice. We have allowed the people who invented schooling to screw up the natural leaning process which is a combination of talking and trying again. That is what engineering has always looked like. It is also laziness that allows a teacher to tell a student they must read.

I happened to be watching a movie from 1943 the other day, called *Princess O'Rourke*. In one scene, women were trying to sign up for volunteer work at the Red Cross. The conversation below was between an obviously uneducated lady who wants to sign up and the Red Cross person who is signing people up:

What would you like to do Anna?
I'd like to learn Red Cross.
Can you read?
No.
It would be very hard Anna. There are things you'd have to study.
You tell me and I will learn.
You would be most useful doing what you know. Do what you can do best.
I can do everything. I have nine children.

Why would she have to study? Because they didn't feel like teaching her. They wanted to make it easy for themselves and have her read. And that is why we have kids read. Because we don't have the time to teach them properly. And since we mostly don't remember what we read, this is absurd. It is time we realized this.

Any good parent talks to their child. They don't answer questions by saying "read this." But school is about mass education and someone has decided that mass education means attempts to make intellectuals for reasons that elude me.

Is there math in engineering? Sure. But it needs to be learned as one needs it. When you need it you can learn it. And you shouldn't be learning from a book anyway. Socrates and I agree on that one:

> Writing, Phaedrus, has this strange quality, and is very like painting; for the creatures of painting stand like living beings, but if one asks them a question, they preserve a solemn silence. And so it is with written words; you might think they spoke as if they had intelligence, but if you question them, wishing to know about their sayings, they always say only one and the same thing.

This obsession with reading and mathematics is just that, a convenient obsession that keeps the test makers happy and makes kids miserable.

There should be a school for the blind that has no reading in it. There could just be talking, and reasoning, and planning, understanding causation, and all the other cognitive processes I have discussed in my book, *Teaching Minds: How Cognitive Science can Save our Schools*. My granddaughter would be just fine in a reasonably designed school system. But we don't have any schools that let kids be kids and learn what they want to learn. We make them study. And that means reading.

One day we will stop texting and stop writing and go back to talking to each other. We will lose nothing from that evolution. We will still tell stories and still learn from experts. And we will do it without reading. There will be plenty of new media. A thousand years from now no one will be able to figure out what all those strange marks were on all the ancient stuff.

Reading is no way to learn

This is a column that attacks reading. No one attacks reading. Let's just assume I am crazy and push on.

Reading is a pretty recent idea in human history. It hasn't worked out. It has given us some pretty good things, like literature, for example, or the possibility of communicating with my audience right now. But these things will be going away soon, and good riddance.

For years, I was an advisor to the Chairman of The Board of Encyclopedia Britannica. My job was to eat dinner with him every few months. At each dinner he asked me if there would still be books in five years. I said that there would be except there wouldn't be his book. "Encyclopedias will disappear," I asserted.

I was thinking about this on a business call the other day. The man I was speaking with was concerned with how training was done at his very large engineering firm. He was rightly worried about "death by PowerPoint." He used an example of learning to change a tire by changing one and then went on to describe quite accurately how we learn in such situations (by practice was the point, something you can't do in Power Point). But, he started his explanation by saying the first step in tire changing would be to get out the instruction manual on how to change a tire and read it.

I said that I had never actually read an instruction manual and that they haven't actually been around for very long in human history. In ancient times, when a young boy wanted to learn to hunt lions he didn't read the instruction manual, nor did he take a class. Throughout human history we have learned by watching someone older than ourselves, trying to copy that person, trying to be part of the team, trying things for yourself, and then asking for help when we have failed. It is not that complicated. This is what learning has always looked like. And then, someone invented instruction manuals and we all forgot what we knew about learning. We replaced human mentors by PowerPoint lectures and asking by reading.

Great. And we wonder why we have trouble teaching people to do complex skills. There is nothing difficult about it. When you need to try to accomplish something that you want to accomplish, you need to have someone who knows how to do those things watch over you and you need to have someone whose work you can observe and copy. You need to be able to try and fail and you need to be able to practice. Reading doesn't come up.

When I say things like this it makes people nuts. The other day I had a conversation with a woman in which I asserted that no learning takes place without conversation. She objected and said that she could look up something in Wikipedia any time she wanted and learn something that way.

No, I said. You can't. She was flabbergasted.

First, let's ask why Wikipedia exists. In part, it exists because Encyclopedia Britannica couldn't keep up. But also, it exists because we live a in a world where we don't know whom to ask. I get asked nearly every day what certain words mean or what certain ideas are about. I am asked because the people I am interacting with know I might know and know that I am always happy to teach. But mostly I am asked because people know that I give quick short answers to their questions. When you have someone to ask, you ask. Reading is the alternative when there is no one to ask.

Let's assume you always had available at your disposal a panel of experts who could be asked any questions you needed to ask. Would you ever read? (That panel is coming soon.) This morning I had a medical question. There was no one to ask. So I started to read. But this is rarely anyone's first alternative.

The second problem with the "I can always look it up" model is simply this - you won't remember what you read. Now we have had a lot of practice at attempting to remember what we read. That practice is called school. We read. We study. We memorize. We take tests. And we are somehow all convinced that we have remembered what we read.

Every year I would ask my students on the first day of class at Yale and Northwestern if they could pass the tests they took last year, right now. No one ever thought they could. They studied. They listened. They memorized. And then they forgot. We don't learn by reading nor do we learn by listening.

We do learn by talking. Assuming we are talking with someone who is more or less our equal and has ideas not identical to ours, we learn by challenging them and ourselves to think hard. We mull ideas. We try out ideas. Even after a good conversation, it is hard to remember what we were talking about. If we do remember it, it means we were changed by that conversation in some way. Something we believed we now have a different perspective on. And we have enabled practice. Practicing talking is like practicing any physical skill. You won't learn to hit a baseball unless you repeatedly hit one over years of practice. The same is true of ideas or facts. Students can temporarily memorize facts but if they don't use them again they will forget them. We need to practice what we know until we are barely aware that we know it, until what we know becomes instinct. We don't know how we talk for example, but we can talk, because we learned how to talk and practice it every day.

Our world has gotten obsessed with reading. Every entrance exam is at least half about reading. People one-up each other by citing what books they have read. If you haven't read one they think is important they can look down on you. (But, it is actually unlikely they remember much from the actual book. They might remember what they were thinking or talking about after reading the book.) This is the modern era. Things have been like this since the invention of texts. Lecturing followed the invention of texts (so the text could be read to you). But this is all going away soon. Socrates noted this in discussing the invention of reading and writing:

> For this invention will produce forgetfulness in the minds of those who learn to use it, because they will not practice their memory. Their trust in writing, produced by external characters which are no part of themselves, will discourage the use of their own memory within them. You have invented an elixir not of memory, but of reminding; and you offer your pupils the appearance of wisdom, not true wisdom, for they will read many things without instruction and will therefore seem to know many things, when they are for the most part ignorant and hard to get along with, since they are not wise, but only appear wise. (Phaedrus 274c-275b)

Reading is going away. Books are going away. There are already better ways of disseminating knowledge. But the schools are difficult to change. Training is difficult to change. People who use the internet can't imagine a life without the tools that are on there now. But there are new tools coming.

The main advantage of reading is that we can skip around. We skim rather than read. It is hard to skim when someone is talking. And then one day, maybe it won't be.

Chapter 4: Technology Saves the World!

Games to the rescue!

Gee, I have a really good idea. Let's make school look like a lot of fun, and have it all on a computer, and let kids play games all the time, while in reality what we will do is make sure students are doing constant test prep. Test scores will go up.

Isn't that a great idea?

It sounds positively nauseating to me but what do I know? That is what is happening. Just a few companies that have come to my attention lately are Enlearn, Center for Game Science, Smarty Ants, Rocket Learning, Renaissance Learning, and Amplify.

Amazingly, every one of these companies has a primary mission to help students meet Common Core standards. (For my non-U.S. readers, Common Core is something bad that is coming to your country soon I am sure.)

Let's put this another way. Bill Gates pushed through Common Core and now is funding companies to make school into an exercise to meet Common Core. In the process, he encourages companies to build software that looks like fun and games but is, in fact, drill and practice on math and reading, all leading to testing to meet Common Core standards.

Here are three such "games":

- *Treefrog Treasure* is a platformer game that teaches whole numbers and fractions as players hop around a variety of worlds.[1]
- *Refraction* focuses on teaching fractions and discovering optimal learning pathways for math education.[2]
- *Creature Capture* is a strategy game that teaches relationships between whole and fractional numbers.[3]

Those sure do sound like fun.

Take a look at their websites and see if you come away with a different conclusion than I did.

Big business has set its sights on making money from education, by insisting on standards and then funding companies that will ensure children can meet those standards.

There is lots of money to be made, and states will be able to announce that test scores are up. School will appear to be a less miserable experience because kids will be playing games on a computer all day. But, of course, what will really be happening is that we will

1 http://centerforgamescience.org/portfolio/treefrog-treasure/
2 http://centerforgamescience.org/portfolio/refraction/
3 http://centerforgamescience.org/portfolio/creature-capture/

produce a generation of children who can pass tests but cannot think clearly, and who have never been taught to think for themselves, plan, diagnose, determine causality, make good judgments, understand the value of something, communicate clearly, or experiment with ideas. But they will be good consumers of the junk being produced by these very same companies.

Congratulations Bill Gates. You have done it again.

But why exactly do you hate children?

Alex Trebek: hero of vocabulary preparation

I was waiting for a football game to come on TV, and there was Alex Trebek selling a vocabulary-building software program. It was one of those infomercials, which was packed with the most amazing garbage about education ever assembled into one half hour. It seems that the company he was touting, Wordsmart, was founded by a "world-renowned educator" named David A. Kay. I thought I knew all the world-renowned educators. Even Google seems to have missed this guy. He sells a piece of software that will not only get your kids great SAT scores and get them into Harvard, but will also guarantee them a high-paying job (not really, they just make it sound that way). (This last nugget is based on the idea that Harvard graduates make more money on average than Joe Schmoe.) And all this will be done through building your child's vocabulary. And why is it important to build your child's vocabulary? Because people who succeed have large vocabularies.

Wow!

I guess people must believe this nonsense, so I checked to see what the software did. Predictably, it tells you a word and then asks you some multiple choice questions about it. It has many ways of doing this, but drill-and-practice is just drill-and-practice by any other name. They are making enough money on this program to buy half-hour spots on national TV. (And they are also able to buy Alex Trebek!)

Now, I assume that most of my regular readers would know why this program is nonsense, but in case you happened upon my writing randomly, here is the point. Just because successful people have large vocabularies does not mean that if you have a large vocabulary you will become successful. Vocabularies are acquired naturally, by speaking to people and by writing to people and by reading, and by interacting with people who have vocabularies a little larger than one's own. This is how we learn words.

This is pretty much the only way to acquire a large vocabulary. You can try to memorize the dictionary if you like, which is more or less what this software is about, but if you don't use the words regularly, you will forget them.

Another piece of nonsense brought to you by those wonderful folks who believe that testing and education are the same thing.

The online learning disaster

The Chronicle of Higher Education has run a few articles lately on the perils of online learning. It seems that students:

Take online courses just to get the credit.

No? Do tell.

Need to be prodded into having discussions online.

Really? Might this be because they are just trying to get the credit?

Don't get the full value of the lecture using this medium.

Actually, this one is beyond my sarcasm. Do people really think lectures are a valid way of teaching? Still?

Have trouble forming "virtual communities."

You mean that nonsense is nonsense?

Find that the quality of education is compromised.

Apparently a lecture hall of 500 students does not compromise educational quality?

And my favorite:

Teachers of online courses find them to be a lot of work.

So, let me set the record straight about online education. It is a disaster. Why? Because every school that offers it has the idea that they will "put their existing courses online." In other words, they will provide the same junk they provide now, but will use a method not suited for that junk. Current lecture courses work for faculty, not students. Students are forced to take them, and they sleep (or text) through them. They take them to get credit, not because they want to, and teachers like them because they are not a lot of work to produce.

But online education will win in the end. It cannot be any other way. This will only happen when what is offered is exciting, experiential, and adapted to the new medium. In other words, it will happen when we rethink education and when we reconsider what it means to teach. Helping somebody do something that they want to do is the right metaphor, not forcing them to endure a set of hurdles in order to get a credit.

Stanford decides to be Wal-Mart

Yesterday I needed a program that my team had written 20 years ago to show to a potential client who needed something similar. The program was done at the Institute for the Learning Sciences (ILS), which I started and ran at Northwestern University 25 years ago. The program was built for the Environmental Protection Agency (EPA), and was about how to run a public meeting. The EPA runs them all the time but they didn't know how to train people to run them. So we created a fictional town with a fictional crisis, and then had the user meet the players (via some very well-done video, acting, and script writing) and eventually start and run a virtual fictional meeting. People they antagonized when they met them individually (virtually) were difficult to deal with in the virtual meeting, which looked, from the user's perspective, a lot like a real meeting.

I thought we really hadn't improved much in the last 20 years. The online possibilities available now hadn't really helped all that much over what we could do then.

I frequently criticize Massive Open Online Courses (MOOCs) in my columns, but that is not my target here. My target is me. Well, not me exactly, but the circumstances that have caused our recent work to be less exciting and real-feeling than the old ILS work.

The reason is, of course, money. The EPA program cost a lot of money to build, and nowadays cheap is the watch-word of online education. Sticking a camera on a professor who is lecturing, or putting a test online is cheap (but a bad idea, then or now). The reason the watch-word of the day is "massive" is that it saves money.

Universities have become Wal-Mart. "We put our courses online." translates to "Maybe now thousands of people will hear our professors' lectures, and imagine how much more money that could bring in. Think of how we can lower prices and sell an even worse product." That is what universities are doing, although they couch it in different terms.

Why Stanford feels the need to become Wal-Mart beats me. But I am sure that Stanford itself won't give the stuff they produce to its own students. No one calls this racism (or classism), but it is education for poor people, just as Wal-Mart is focused on poor people. Stanford students won't eat what Stanford sells to others, but it is selling it like mad, to those folks who will never see Palo Alto and will never access a real Stanford education.

What they are selling is not very good. It certainly isn't as good as what we could do 20 years ago on a computer, when there was money available to invent new kinds of education to teach real skills. The money spent by venture capitalists, out to build the next

Facebook or Twitter of education, is meant to make money, nothing more and nothing less. Why Stanford isn't ashamed of itself I don't know.

Wal-Mart isn't ashamed of itself, because it provides low-cost stuff to people who can afford only that. Stanford provides high-cost stuff to the elite of the elite, so one can only guess why they want to become Wal-Mart.

As for me, I would love to go back to the old days (at ILS), when money wasn't the issue, and quality was. My people still produce quality. We have learned to improve on what we did before (more mentoring, less multiple choice; more teamwork, less one-on-one with the computer), and to get by on less money by inventing powerful tools.

Still, I long for the days when we weren't competing with Wal-Mart. Actually, I apologize for the analogy. Wal-Mart was my client in those days, and they wanted real high-quality training in a virtual world for their employees too.

Greed is not disruption

The *New York Times* ran one of their nuttier articles about online education yesterday, this time about Harvard Business School's (HBS) new online plans.

> Universities across the country are wrestling with the same question — call it the educator's quandary — of whether to plunge into the rapidly growing realm of online teaching, at the risk of devaluing the on-campus education for which students pay tens of thousands of dollars, or to stand pat at the risk of being left behind.

The truth is somewhat different. I talked last week with a university president I have known for many years, and asked him why he was building online courses. His answer, unsurprisingly, was "fear."

A few days later I met with the online division of a very well-known university, and asked them the same question. Their answer was that test scores improved in courses where students had access to the lectures online. The real answer, I think, was that they were also afraid, and that some Foundation had given them money to do it, so they did it.

The *Times* goes on to quote Clayton Christensen, a well-known Harvard Business School professor:

> He said he remembered listening to an edX presentation at an all-university meeting. "I must confess I was unsure what we'd be really hoping to gain from it," he said. "My own early imagination was: 'This is for people who do lectures. We don't do lectures, so this is not for us.'" In the case method, concepts aren't taught directly, but induced through student discussion of real-world business problems that professors guide with carefully chosen questions.

This is, of course, the actual problem. If HBS or anyone else wants to build online curricula, the question they should be asking is: what exactly are they putting online? Courses followed by tests? Really? This, according to the *Times*, is what HBS is doing:

> Students have nine weeks to complete all three courses, and tuition is $1,500. Only those with a high level of class participation will be invited to take a three-hour final exam at a testing center.

If this is what they are doing, then they should be ashamed. People go to HBS so they can say they have a Harvard MBA. I was a professor for way too long to still believe that students are there primarily to learn. They want credentials. HBS, to its credit, has typically offered courses that involve argument and discussion, not tests. Online lectures, followed by tests, are a parody of what real learning looks like.

The university world has lost its collective mind. Fear (and greed) has driven them to take their worst educational devices, lectures and tests, and try to make that the cornerstone of the future.

So, let me say it one more time:

- We learn by doing, not by listening.
- We learn by mutual story exchange in a conversation.
- We learn when we have a goal, something we are trying to accomplish, and by that I do not mean gaining a credit towards a degree.
- We learn when we have peers and mentors with whom to discuss the things we are working on.

MOOCs (and lectures in general) pervert what it means to teach. Teaching isn't telling – it actually involves listening, helping, suggesting, and so on. Universities know how to do this. That is how most PhD programs work. Massive education is not about learning and it never was. Yes, professors like lecturing. I like lecturing. I just don't delude myself that my lecturing is teaching anybody anything. When I want to teach somebody something, it involves constant interaction. By interaction I do not mean stopping a video lecture and guessing what comes next (which is what the Wharton Business School seems to be doing).

It would be nice if all the universities really did want to build online courses because they were trying to be disruptive (to use Christensen's word). But universities are very afraid, and have never seen disruption as their goal.

Some will, however, and those that do will succeed by providing something other than lectures and tests.

The system will change soon enough. I doubt HBS will lead the way, but there are universities out there who intend to do just that.

The computer is a powerful device for doing. Time to get busy and use it that way folks.

Princeton Professor teaches Coursera course; you must be kidding me!

I don't recall ever agreeing with anything Thomas Friedman has ever written in the *New York Times*, but this Sunday's article was especially ridiculous.

He was again extolling the glories of the coming education revolution led by MOOCs. This is part of what he wrote:

> Mitch Duneier, a Princeton sociology professor, wrote: "A few months ago, 40,000 students from 113 countries arrived here via the Internet to take a free course in introductory sociology. ... My opening discussion of C. Wright Mills's classic 1959 book, 'The Sociological Imagination,' was a close reading of the text, in which I reviewed a key chapter line by line. I asked students to follow along in their own copies, as I do in the lecture hall. When I give this lecture on the Princeton campus, I usually receive a few penetrating questions. In this case, however, within a few hours of posting the online version, the course forums came alive with hundreds of comments and questions. Several days later there were thousands. ... Within three weeks I had received more feedback on my sociological ideas than I had in a career of teaching, which significantly influenced each of my subsequent lectures and seminars."

Friedman mentions this because he thinks it is a wonderful thing, I suppose. Let's consider what this professor actually said:

> My opening discussion of C. Wright Mills's classic 1959 book, 'The Sociological Imagination,' was a close reading of the text, in which I reviewed a key chapter line by line.

Well, isn't that just education at its finest? Princeton should be proud. Not only are they still lecturing, a relic of the Middle Ages when students didn't have books and monks read them aloud, but the professor is reading it line-by-line. The analysis of a text is a scholarly activity done by intellectuals, and when done with students, it is part of an effort to create more intellectuals. Does the professor think that the world needs 40,000 more sociology intellectuals? When this stuff happens at Princeton, it still isn't really good educational practice, but Princeton does try to produce intellectuals for the most part.

When done with 40,000 students from 113 countries, this is simply fraud. There is no need for them to read a text in this way. Far from being a revolutionary new practice that will eliminate universities as Friedman says, this kind of activity is perpetuating the very thing that is wrong with universities – their distance from the real world.

> ...within a few hours of posting the online version, the course forums came alive with hundreds of comments and questions. Several days later there were thousands. ...

It is nice that there were thousands of comments. How many did you respond to, Professor Duneier?

I assume the answer is "none." As a professor, not responding to a student is, in my mind, the worst thing one can do. Education is about the dialogue between professor and student. This is why classrooms, especially large classrooms, are a terrible idea. They limit discussion. When I taught at Yale and Northwestern I never assigned readings, just topics for discussion. And then we discussed. If you had 30 or 40 students, you could get into some good arguments, especially if I had assigned a provocative question to think about. ("What does it mean to learn" was one I often used.) Your job as a professor is not to notice how many nice discussions students have with each other.

It was the last line that really got me:

> Within three weeks I had received more feedback on my sociological ideas than I had in a career of teaching, which significantly influenced each of my subsequent lectures and seminars.

So, the Coursera experience was good for you, eh? Nice to hear. But the issue is that universities have always been good for the faculty. Places like Princeton are run by the faculty, for the faculty. No one teaches much. No one cares about anything but PhD students and research. Undergraduates sit in lecture halls in order to pass the time between football games and parties. No one cares because they all wind up with impressive Princeton degrees.

Friedman is right that online teaching will change universities, but not the kind of online teaching that Coursera provides.

Just yesterday, there were thousands of visits to a lecture of mine that is online because it was assigned as part of a Coursera course. I find that very funny, since my lecture was about why lectures don't work (oh the irony!) and why learning requires doing and why universities should stop teaching scholarly subjects and start teaching students skills they can use in real life.

Yes, change is coming. Too bad Mr. Friedman doesn't have a clue why. Here it is. We can build mentored learn-by-doing courses online that challenge current teaching practice. They won't be offered by Princeton because Princeton likes what it has now. But change is coming, just not the change Coursera or Friedman have in mind.

Efficiency is not reform

My attention was drawn to a blog post written by a very well-respected professor at Columbia University named Jeffrey Sachs. In it, he asserts that productivity is improving in our society, and cites the following as evidence of this in education:[4]

- At eight on Tuesday mornings, we turn on a computer at Columbia University and join in a "global classroom" with 20 other campuses around the world. A professor or a development expert somewhere gives a talk, and many hundreds of students listen in through videoconferencing.
- At Stanford University this fall, two computer-science professors put their courses online for students anywhere in the world; now they have an enrollment of 58,000.

I found these supposed pieces of evidence astonishing in their naïveté. Of course the man is an economist and not someone who thinks much about education, one would assume. But still.

I have often said that the main problem in fixing education is the professors. "We have met the enemy and it is us" applies very well to why education is so hard to reform.

Really, Professor Sachs? You are excited by that fact that more people can listen to your lectures? Ask any college student what he can recall from a lecture an hour after he has listened, and see how much he remembers and how much he simply remembers wrong. Lecturing is a completely archaic way of teaching. It exists today at top universities only because hot-shot professors at top universities (of which I was one) think their time is better spent doing almost anything other than teaching. Talking for three hours a week seems like a pretty good deal, enabling them to go back to doing what they really like. No one learns in a lecture. If you cared about education you would stop lecturing. But you care more about research. This is fine – I did too when I was a professor. But recognize that you are the problem in education and that video conferencing is the solution to nothing.

Sachs makes the same point twice when he cites the Stanford course. The Stanford online Artificial Intelligence (AI) course has gotten a lot of media attention. AI is my field (and one of the instructors was a PhD student of a PhD student of mine). I don't know what is in the course and I don't care. The media doesn't care either, nor does Sachs. They just like the 58,000 number. What if I said that a former student of mine was a great parent and so he was now raising 58,000 children online? Would anyone think that was a good idea? This may seem like a silly analogy unless you really think about it.

4 Services without Tears http://www.project-syndicate.org/commentary/services-without-tears

Teaching, as I point out in my book *Teaching Minds: How Cognitive Science can Save our Schools*, is basically a one-on-one affair and is about opening new worlds to students and helping them do things in that world. This will not happen in a 58,000-student course any more than it happens in a 100-student course. Lecture courses are just rites of passage that we force students to endure so they can eventually start working with a good professor in a closer relationship (at least, this is what happens at a good university). A book would do as well for this, and better would be a well-constructed learning-by-doing online course.

But what is happening in today's world is that the action in educational change is all about getting bigger numbers online without trying to improve quality. Stanford is making a lot of noise with this course but nothing good can come from it.

Professors need to stop and really think about education. Of course, the problem is that they have no motivation to do so. They are well paid and having a good time. Only the students suffer.

MOOCs, the XPRIZE, and other things that will never change education

I have been in computer science nearly all my life, but I didn't get interested in education until the early 80s. I looked at what was going on in computers and education work at the time, and was unimpressed. The people involved seemed mostly interested in programs that helped kids do algebra better, or taping and sending out lectures in what was called "distance learning."

I found these things totally uninteresting and began to think about how to really use the power of the computer to build simulations, games, or activities that would excite kids about learning. But, it was hard to get schools or publishers interested in what I was doing, because they were committed to making no changes whatsoever in how education had functioned for the last 1,000 years.

Well, that was 35 years ago. What has changed? Nothing, it seems. There was the MOOC craze, which is a different way of taping lectures. Fortunately this craze seems to be over.

In *The Chronicle of Higher Education* today there was the headline "Optimism About MOOCs Fades in Campus IT Offices."[5]

So, that's nice. Nothing good will happen online for a while because of MOOCs, but we can stop pretending that education is about listening to lectures and passing tests, and go back to understanding that real learning has always been about – trying to do something you want to do, and having someone available who knows more than you and is willing to help you do it. (Sometimes these are called teachers.)

But then I read a headline in the *Washington Post*: 'New $15 million Global Learning XPRIZE wants to disrupt education as we know it."[6]

Wow! Great! Someone with real money wants to change education. Oh, wait. Too soon to get excited. This is what they are worried about:

> As Diamandis emphasizes, what's needed is a new way of thinking about education if we plan to educate tens of millions, if not hundreds of millions, of children: "Traditional models of learning are not scalable," Diamandis said. "We simply

5 http://chronicle.com/blogs/wiredcampus/optimism-about-moocs-fades-in-campus-it-offices-survey-finds/54705

6 http://www.washingtonpost.com/blogs/innovations/wp/2014/09/22/new-15-million-global-learning-xprize-wants-to-disrupt-education-as-we-know-it/

> cannot build enough schools or train enough teachers, which brings us to a pivotal moment where an alternative, radical approach is necessary."

So, traditional models of education (kids packed into classrooms, all doing the same stuff despite their interests, and being made to pass tests and listen to lectures) are not scalable! Right. That was the problem. If only we could have more of that stuff that has made children miserable for 2,000 years. (Petronius Arbiter said in the year 60, "I believe that school makes complete fools of our young men, because they see and hear nothing of ordinary life there.")

And, there won't be any more teachers if the XPRIZE has its way:

> ...there won't be teachers the way we think of teachers today. Even students learning autonomously will require much more peer-to-peer learning, in which students armed with apps and tablets teach each other about the world. Finally, there won't necessarily be "courses" or "learning modules" involved in the next iteration of educational innovation. There will be software and apps, and it will be up to the prize teams to define exactly what these do.

Their intention, as I understand it, is to eliminate the only things that matter in learning. These are, in my opinion:

1. A goal
2. A person who can help you more clearly define that goal
3. People to work with towards that goal
4. New goals that come from having accomplished that goal
5. Being able to fail and get help
6. Being able to write and talk about what you have done
7. Fielding reactions from those who can help you improve what you have done
8. Help in thinking about what else you might accomplish

Instead, they want kids on tablets using apps to learn the same old crap we have always been teaching, but this time they are on their own. Yeah!

I believe that learning is a conversation, as I have said in this space before. Technology is helpful to the extent that it lets you try things you might not have been capable of doing before. Design an airplane, start a business, plan a career, invent something. These require teachers (or mentors), and plans of attack, and simulations, and expert advice. I still believe that technology can save education, but we need to define more reasonable goals.

XPRIZE chairman Peter Diamandis says:

> We're aiming at kids who live in villages where there's nothing. This has to take them from complete illiteracy to basic reading, writing and numeracy.

Or to put this another way, yet again, someone is trying to teach the same old junk. There is nothing wrong with learning reading, writing, and arithmetic. There is a problem with every subject taken after third grade, however. The only thing that is good about the first three grades (apart from the 3 R's) is the presence of the teacher. And the teacher is what they want to eliminate.

Here is an idea. The best use of technology in today's world would be to hook up those who want to learn with someone who wants to teach them. You could learn a language online by talking on Skype to someone who speaks that language. You could learn how to start a business by talking with business experts and discussing your plans. You could learn to discuss complex ideas in a Socratic seminar lead by.... a teacher. The real value of the computer in today's world is that everyone is connected. And there are a lot of experts (people my age who have little to do, for example) who would be happy to mentor kids who wanted to do things that were not part of their existing school systems. The XPRIZE, I assume, will soon have a prize for the computer that tutors algebra most efficiently, and we will be back to where we started.

MOOCs: The New York Times gets it wrong again; Europe is not lagging behind the U.S.

Again I feel compelled to write about MOOCs, but this time because of a recent *New York Times* article entitled "Europeans Take a More Cautious Approach Toward Online Courses."[7]

The *Times* has been thrilled lately with MOOCs as a topic, but has shown no interest in addressing the real education issues in its discussions of them. Take, for example, this quote from Adrian Smith of the University of London, printed in the *Times* article:

> "However, you ignore them at your peril," he said. "The challenges they pose to the traditional classroom model of knowledge transmission are obvious. The question is no longer whether we should consider MOOCs, but how quickly to get involved."

According to Mr. Smith and, one would assume, to the *Times*, MOOCs challenge the traditional classroom model in some important way. But how do they do that? The very idea of education as "knowledge transmission" is the problem in the first place. Yes, MOOCs transmit more knowledge to more people, but knowledge transmission as education is exactly what is wrong with the standard university education model, and most education of any sort.

The idea that people can learn by being told has been ridiculed both by movie-makers and by philosophers (such as John Dewey and Immanuel Kant, not to mention Plato). Still, we persist. Do you know how many life rafts there are on a 757 and where they are located? Why not? That knowledge has been transmitted to you, tens or even hundreds of times, every time you board a 757. No one listens. Still we talk.

The *Times* quotes John Zvereff from Barcelona's Open University:

> "I think we should be responding to challenges to the learning model," Mr. Zvereff said. "At U.O.C., our philosophy is based on accompaniment — each student is assigned a tutor who stays with him or her all the way through till graduation. None of the MOOCs I've seen offer anything like that."

What a weird idea. Teachers are important. Yet, teachers are exactly what MOOCs don't have. The valid part of the classroom model is not the lecturing or the tests. It is the idea that there is a teacher present who you might engage with in some intellectually stimulating way. The idea of school, at least in principle, is the interchange of ideas that

7 http://www.nytimes.com/2013/02/18/world/europe/18iht-educside18.html?smid=tw-share&_r=0

cause you to challenge cherished assumptions and grow intellectually. This requires a good teacher. That is the teacher's job, not knowledge transmission. MOOCs pretend to do this of course by having students talk to each other, which is probably as useful as all the comment section after a *New York Times* article. If the *Times* let readers interact with the authors of its articles, and it was the author's job to respond to every comment, then maybe the *Times* would show that it knew something about teaching. Perhaps the *Times* likes MOOCs so much because they are so much like newspapers – great at the one-way transmission of knowledge.

Another quote from the article, this time from Drummond Bone, master of Balliol College at Oxford, who:

> ...told the conference that his university "was very strongly associated with face-to-face, and very expensive, one-to-one learning," and was unlikely to change.

Good for him.

But, here is the most honest quote from the article:

> Veronica Campbell, the dean of graduate studies at Trinity College Dublin, said her school had no open online courses, nor plans to offer them, but "there is a fear of being left behind, so we are considering what to do."

This is the truth of the matter. Universities everywhere, not just in Europe, are afraid. The *Times* and others are promoting MOOCs, so universities are afraid of the MOOC tidal wave (for which the *Times* is, in part, responsible).

But this is what universities are actually afraid of:

- High quality education, the face-to-face, challenging, interactive type, is very expensive to produce. Even the best universities offer mostly lectures. The reason is that small seminars taught by famous researchers are too expensive to run often. They must be balanced by lecture halls seating a thousand students, even though almost everyone knows this is not real education.
- Online learning will clearly beat classroom learning eventually. By online learning I do not mean MOOCs. Why will online win? Because one can produce authentic experiences, where students learn by doing, with mentors available to help them produce things, at a lower cost than it takes to run a giant campus. Build the best Search Engine Optimization course, or Data Analytics course, for example, and no university ever has to put up the money to build another. They just have to provide mentors for that course.

By the way, Europe is not lagging behind at all. A new open university offering online learn-by-doing has just been launched in Andorra.[8] They offer an experiential MBA that I built for them. No U.S. university would dare offer anything like it. Why not? Because universities in the U.S. are run by their faculties, and faculties want no challenges to

8 La Salle Open University http://www.uols.org/en

the existing model. A faculty member may be happy to be the lecturer in a MOOC just for fun, but if that became their actual job they would be running for the hills. Faculty control of U.S. universities means they will never change, since faculty members have a pretty cushy life that they want to protect.

This is different in other countries, where the best university education is not private. The *Times* needs to understand that the U.S. is not the center of the university world. It may have that position in research, but not necessarily in teaching. And other countries may well be willing to change a broken teaching model that relies on knowledge transmission, because their citizens are demanding it.

Why are universities so afraid of online education?

A climate of fear is enveloping our major universities. One after the other they are joining well-capitalized, venture-financed operations that offer free online courses. The companies are paying the universities, so of course the universities are taking the money. What do they have to lose?

New program offerings appear regularly, the latest being one that wants students to attend classes remotely and pay full tuition for the privilege of doing this.

Something important is going on, but it is not quite obvious what. Well, it is to me.

The universities are desperately afraid. Of what?

The university that started all this was MIT, when it announced over a decade ago that it would put all its course materials online, free for all to use. The press made quite a fuss about this, but I said at the time that MIT just wanted to appear to be doing something. It knew very well that the course materials that professors prepare constitute a very unimportant part of what it means to receive an MIT education. (What is important at MIT? Working with faculty and students to create new ideas and new projects.)

I was asked if I wanted to head up that operation, and I told MIT that I would make real course offerings to create a worldwide MIT online delivery system. I was never called back.

A decade ago I built a series of online masters degrees for Carnegie Mellon University, and was not only not praised for doing this but was also immediately fired. I was explicitly told that Carnegie Mellon didn't want to sully its brand by having too many Carnegie Mellon degrees out there. They want to be an elite brand name, as do all the major universities.

But suddenly it seems that the game has changed and every university wants to go online. But, this is not really the case.

To understand this, you have to think for a moment about courses and what they are all about. Most students take four or five courses at a time as full-time students at a university. While they are doing this they play football or work for the student newspaper, or maybe even hold down a real job. Plus, there are a great many social events to attend, in addition to the constant action of dormitory life.

In the life of your average college student, a lecture course is something to be barely paid attention to at best, or slept through at worst. The fact that a friend can make a video recording of lectures for you means you can skip them altogether.

And this, of course, is the origin of online courses. As long as someone is making a recording of the lecture available to their friends, why shouldn't the university do that, and say it was what they wanted to do in the first place? Add a quiz or two, and no one ever has to show up. Voilà! Coursera!

But why do the universities agree to this? The answer, as always, seems to be money.

But really, the answer is fear. It is important to understand what exactly they are afraid of.

Here are four things universities are deathly afraid of:

1. What if the model that "everyone must go to college" stops being pushed by employers and governments?
2. What if they simply can no longer charge large tuition fees to students?
3. What if professors, who at top universities are primarily researchers, were actually made to teach as their primary activity?
4. What if the students stop showing up on campus?

The money issue is a big one. Tuition amounts have risen way ahead of inflation, supported by readily available student loan programs and by the belief that anyone who doesn't go to college is more or less useless. We fail to observe how many successful people have never graduated college, including Bill Gates, who never stops promoting school standards, teacher evaluations, and now online courses. Mr. Obama wants everyone to go college, as do the authorities in the U.K. Why exactly? Because the universities are afraid, and they are lobbying hard for this. When you need a PhD to work in McDonalds, however, the model will fall apart, and we are headed in that direction.

All that tuition revenue, and donations from alumni who fondly remember the great football teams and parties, help sustain what is actually an absurd model, and every university knows and fears the downfall of that model.

The model is what I like to call the "superstar system." Top universities compete for superstars in the same way that baseball teams and movie producers do. There are only so many big names, and the university that has the most wins. If Harvard has more Nobel Prize winners than Yale, Yale is thinking about this all the time. (I say this as someone who was on the Yale faculty for fifteen years.)

Research universities want to sustain the model that has made them great places to live and work. I loved working at them for 35 years. But the students were, and are, being cheated. Some professors care about the undergraduates at an Ivy League school, I am sure. I certainly didn't.

I was once yelled at by an undergraduate who said he had paid big tuition to Yale and that I should meet with him at times other than my few-and-far-between office hours. Of course he was right. But the incentives at the research universities are all about publishing and international fame, not about happy undergraduates.

(I did meet with him, by the way, and he eventually became a researcher at a major university where the undergraduates have great difficulty finding him.)

Just the other day, Northwestern University, where I ran the Institute for the Learning Sciences for many years, announced proudly that they would let people attend classes remotely if they met admission standards and paid full tuition. They should be ashamed of themselves. There are still plenty of people at Northwestern who know how to do online education correctly. We pretty much invented it there.

But, what we invented was using the computer as a learn-by-doing device, eliminating lectures and classrooms, and replacing them by projects one could do on the computer with the help of faculty and other students.

I am slowly discovering universities who want to use this model online, but the faculties always object to it. No lectures? No theories? Just learning by doing? Oh, the horror. The faculty might have to teach.

So, don't be too impressed by MOOCs, non-MOOCs or any other nonsense that keeps 'courses with a teacher talking' the staple of university education. Students put up with that because they get degrees they can brag about, not because of all the wonderful stuff they learned. It is not any more wonderful if you are at home in your pajamas.

More online nonsense: Starbucks and Arizona State agree to do nothing useful

In a deal that was announced with great fanfare, Arizona State University and Starbucks agreed to let baristas get online degrees at ASU. The facts were a little different from the way they were first perceived. This is from *The Huffington Post*:[9]

> Arizona State University President Michael Crow told *The Chronicle of Higher Education* that Starbucks is not contributing any money toward the scholarship portion. Instead, Arizona State will essentially charge workers less than the sticker price for online tuition.

Still, this is certainly wonderful, because now Starbucks employees can finish their degrees. Oh. Wait. This is online education, so it will be terrible.

Here's my favorite part from ASU about their online degrees:[10]

> In this course, the activities for the week include discussion posts, readings, audio narrated slide lectures, e-text content, a work sheet activity, and web links, a podcast, and videos.

The video says that all ASU online courses are like this. In other words, students get to listen to lectures, read, and post their points of view. After this exciting educational experience, they will have earned credits, and after enough credits they will earn a degree. As for getting an education, well, not so much. They will probably be qualified to be baristas.

We fail to recognize that online programs are typically unimportant and deceitful. Online programs that will get you a degree, when you have never actually done anything but read, listen, post to a discussion board and then take a test, are simply not actual education.

So ASU has joined as another player in the "we don't take learning seriously" market.

Learning, I will mention one more time, involves doing, which involves trying and failing, and is best done under the guidance of something called a teacher (who helps you improve your work).

---

9 Starbucks Isn't Spending As Much As It Wants You To Think For ASU Scholarships http://www.huffingtonpost.com/2014/06/19/starbucks-asu-scholarships-spending_n_5512376.html

10 ASU video: http://asuonline.asu.edu/how-it-works/learning-online-at-asu

If a computer and the web are to be involved, we need to build Mentored Simulated Experiences, where doing actually takes place (and where there are no lectures).

I can understand why Starbucks doesn't care about actual education for its employees, but I know Michael Crow and I thought he did care about education.

I guess I was wrong.

Students: Be very afraid of online degree programs, especially if Pearson had anything to do with them

The other day I read this article in *Politico*, "No profit left behind. In the high-stakes world of American education, Pearson makes money even when its results don't measure up."[11]

Anyone who cares about education knows that Pearson is running our school systems through its tests, grading of tests, and nearly anything else it can think of. What I learned in this article is that they are now a major provider of online courses to universities and virtual high schools as well. So, I thought I would take a look at their courses which would of course, be the usual crap mixture of reading and taking tests interrupted by a lecture.

But, the University of Florida, one place that buys these courses, doesn't show them to the casual viewer. Instead it provides a promo video which includes the following keywords:

innovative pedagogy
positive game play
social interaction
promotes pure learning
encourages collaboration
values individual student identity
builds upon student's strengths and interests
allows for student choice
provides opportunities for reflection

If I didn't know better, I would have thought they were describing the learn by doing, experiential, mentored simulation courses that my team and I have been building for over a decade. We too have created an innovative pedagogy that encourages (in fact, it requires) collaboration, provides choices, and is big on reflection.

---

11 No profit left behind http://www.politico.com/story/2015/02/pearson-education-115026.html#ixzz3RjPrnPkH

I don't know what *pure learning* is and I don't know what it means to *value student identity*, nor do I think *positive game play* means anything, but, I am getting the idea that Pearson (and Florida) are learning innovative educational vocabulary rather then learning how to build good educational experiences.

I watched more of the University of Florida video. I learned that, "students interact with each other."

This, I have found, means that there are student discussion boards and students get to talk to other students just as they do in MOOCs without ever actually involving the faculty in any way. There are "team discussions" which would be fine if they ended up in something shared and engaged with by a teacher, but that doesn't happen. What does happen is the use of "playful promo videos for each module topic," which seems to mean a funny intro of some sort. There are also short "weekly constitutionals covering foundational topics." I have no idea what that means. Maybe it means the lectures they assure us they don't use or the readings they don't mention they will make you read.

They do have an instructor however:

> The instructor interacts with students via Twitter, live tweeting during public events, and sharing content related to course activities.

Wow. The students get to read tweets from a professor. Now that is an innovative pedagogy!

We also find that:

> The course does not rely on assigned readings and multiple-choice assessments (although all those are featured to a limited degree).

I have no idea what that sentence means, but my best guess is that course is nearly all assigned readings and multiple choice tests since that is what Pearson does for a living and that is what Pearson is ramming down the throats of every student online and off line whenever it can do it.

We are told that students complete missions. What is a mission you ask?

> Missions are the experiential component of the course: They have to interview people, they have to talk to people, they have to do research and they have to build something, whether its something as simple as an essay or maybe even an infographic, a digital timeline, or a video.

So students are writing essays as usual, but they can also make graphical or video essays. I wonder who looks at them. The twitter bird?

So this is what I learned: Be very afraid of online courses. They are worse than live courses by a lot, and live courses are usually just boring lectures and tests.

Students: Be very afraid of these online degree programs because if Pearson continues to be in charge they won't be worth the price of printing the diploma. You will have learned nothing except how to argue with other students on a discussion board and take lots of test and complete many "missions."

The University of Florida may use the vocabulary of experiential learning but accomplishing real live tasks, tasks that someone might one day actually employ you to do, requires the learning and practice of real skills. But, building courses that simulate actual experiences is expensive, and neither Florida nor Pearson is willing to spend much money on building new things. If you want to see what an experiential learning by doing course should look like, take a look at the Experiential Teaching Online (XTOL) course offered by University of Texas at Austin, modeled on my story-centered curriculum approach.[12]

They can steal our vocabulary, but they can't copy what we do, mostly because they really don't want to.

Universities are, for the most part, not concerned with teaching. I also watched the video promo of an online University of Florida Psychology course where the speaker was the instructor. She never said what the online course was like but she did say the word "research" about ten times.

12 http://www.xtolcorp.com/courses/see-a-demo/

What online education should be

I learned something yesterday. (This is not an everyday experience at my age.) I met with a group of faculty from a university that was thinking about adopting some of my online learn-by-doing curricula. I don't typically meet with faculty about this because, in general, faculty don't care about educational change and they aren't the decision makers anyway. I learned that I had been invited to talk with the faculty to allay their fears about online education.

I hadn't really thought about this before. Of course, I know that faculty at places like San Jose State are objecting to MOOCs for a valid reason. MOOCs are providing canned lectures that are essentially faculty-job-eliminators. Stanford may be pushing MOOCs, but they surely won't be using them much. Faculty members need to lecture in order to pay the bills. At places like Stanford, the faculty care about research and... did I mention research? They should be happy to not have to lecture. But if they don't, who will pay their salaries? Some superstars can pay their own salaries from their research funds, but the average faculty member is actually being paid to teach, despite the fact that they get no respect for it and often do it badly. Stanford will muddle on and will be around for a long time. Not so San Jose State, which could easily disappear if state officials widely adopt MOOCs.

So at the meeting, which was with a good, but hardly Ivy League private college, the faculty were afraid. I didn't realize what exactly they were afraid of for a while.

It was me.

They were afraid of me. They were the kind of faculty that dominates the educational landscape, but not the kind of faculty that I have encountered in my professorial life at Stanford, Yale, and Northwestern. While research dominates the lives of the main faculties I have encountered, teaching dominates the life of the faculty with whom I was talking yesterday.

They were worried that if their university adopted my online masters degree programs, they would lose their jobs. After listening to me talk for a while (I was still, at this point, oblivious to their concerns), they started making odd statements like:

So you think that the problem at universities is that there isn't enough good teaching?

Yes.

Your online courses wouldn't take away our jobs?

Hardly, we would need you to supervise the new mentors you would need to hire.

And my last and favorite:

Isn't this the way people have always learned, and the way universities always used to teach?

Yes. Exactly. Mother chimps teach their children by showing them what to do and then helping them do it. Professors teach PhD students one-on-one, supervising their work as they try things out. No one gives lectures to their children at home.

Online learning, in my mind at least, was always supposed to make learning more fun and more relevant to each particular student, and was meant to require a heavy investment from faculty to improve the nature of the university through better teaching. This does not necessarily mean more teaching, but rather individualizing teaching in a way that online learning makes possible.

So this is another thing MOOCs have screwed up. They have put faculty in fear of losing their jobs (rightly so) when the real issue is how to use online learning to improve the teacher-student experience.

Here is a picture of me doing one-on-one teaching with my grandson. I am teaching him how to handicap a horse race.

Chapter 5: Everyone Should Go To College?

Everyone must go to college. Does anyone ever ask why?

I've written about why high school is a waste of time. College is not exactly a waste of time. Some people make very good use of college, and for many it is a lot of fun. But the idea that everyone must go to college is simply wrong. It is constantly reinforced by politicians, and as long as employers insist on only hiring college graduates, they may be right. But the general public has many illusions about what goes on at college.

The top ten good things about going to college

1. When you graduate, employers will think you are now more employable with a degree.
2. There will be lots of very good parties.
3. You will make some life-long friends there.
4. There is a world of knowledge to which you will be exposed.
5. You may meet some very smart professors who may have some time for you.
6. You will have great conversations long into the night with your dorm mates.
7. If you attend college away from home, living on your own will make you grow up.
8. You may learn to talk like an intellectual.
9. You will have fun.
10. You will try things (some not so wise) that you never tried before.

You may notice that I failed to mention much about education in this list.

The top five illusions about college

1. There will be great courses

Well, not so many really. Most courses meet for three hours a week, and most are lecture courses. You really can't learn much in three hours per week, and it is almost impossible to remember a lecture a week after you heard it. Why do courses meet for just three hours per week? It is very convenient for professors. That way they do not have to teach too much. At our top research universities (where I worked for 35 years), research is much more important than teaching ever is, and a division of three hours of teaching and 37 hours of research seems about right to professors. I can assure you it seemed just fine to me. I brought in research money and I didn't have to teach much. That is the deal at the top universities. It is a good deal for everyone except the undergraduates.

2. I will major in something I love
The idea of majors was not put in place for the benefit of undergraduates. Majors serve a purpose for research-oriented faculty. They make students concentrate on an area so they can more quickly be herded into the advanced research courses that are the only courses research-driven professors actually want to teach. They also enable departments to require courses that no student would ever want to take. These are typically courses that are very unattractive to students but very important for faculty, because otherwise no one would sign up for them and those faculty would have to teach introductory courses, which no one ever wants to teach.

3. I will be employable with my college degree
Not if you major in English, history, political science, linguistics, mathematics, physics etc. Employers know that undergraduates have simply taken a smorgasbord of courses and have very rarely learned anything much at all. Big companies hire college graduates and immediately start training them to do what that company does. No one expects undergraduates to actually know anything at all. It is an unwritten bargain: if you want to work in a big company, just get good grades, then we will know that you will do what you are told. The companies will figure out what to teach you after you graduate.

4. I will be better off at an Ivy League school
This country has maybe 25 or 50 top research universities. The Ivy League has eight of them, and there are many others. The students are smart there, they work hard for the most part, and they take life seriously. But Yale (just an example because it is the school I know best) has a mission that its students don't know about. It is trying to train professors. Every research-driven professor (and that is whom Yale always aims to hire) wants to steer their undergraduates into their line of research. This is certainly what I did and it is what every faculty member wants to do. So, if you want to have a research career, Yale is the place for you. But what if you don't care about research? Why spend all that money? There are plenty of other colleges.

5. There are hundreds of good colleges in the U.S.
Well, maybe not. The schools that are not in the top 50 want desperately to make it into the top 50. So even those ranked in the bottom thousands want very much to be research universities, and brag on their websites about the great research going on. Why an undergraduate would care about research unless he or she wants to be a researcher is beyond me. But the people who run universities don't care about that. You never hear them advertising "come here and we will get you a job."

There is no easy answer to all this. Universities will not change any time soon. They have no reason to.

Should I go to a "hot" college?

The *Daily Beast* posted a list of "hot colleges" the other day, which reminded me of exactly how insane this country has become about going to college. It is actually quite difficult to choose what college to attend. But, as a retired professor, I find the concept of a "hot college" rather amusing. I can recall, when I was working at Yale, that every now and then Brown would become determined to be hotter than Yale. It was hard to fathom what this might mean. We had more or less the same faculty we had the previous year, as did Brown. The quality of students was more or less the same at both schools. The campus hadn't changed. How did Brown get hot, and then, later, get less hot?

When I was working at Northwestern, one year we were suddenly "hot." This time I knew why. Our football team had played in the Rose Bowl the year before. This is, of course, a very clever way to choose a college – by examining the quality of its football team.

As a professor, one is aware of other faculty in one's field, and in allied fields all across the world. Ask any professor about another university and he will judge the quality of that school by the quality of the faculty he knows or has heard of. This is not a bad measure, although it is an idiosyncratic one. Thus, I was surprised to find, on the *Daily Beast's* "Top 15" list, some schools that I had either never heard of or could not name a single faculty member, namely Elon University, University of Georgia, Washington and Lee, Ohio Wesleyan, and University of St. Andrews.

I have no ability to judge the quality of these schools, nor do I have any interest in disparaging them. I am concerned instead with the folly surrounding college entrance and college choice. So with that in mind, what makes these schools "hot"?

According to the *Daily Beast*:

- Elon is hot because it "has gone out of its way to recruit applicants interested in the sciences by luring them with the possibility of undergraduate research."
- Ohio Wesleyan University is hot because "Loren Pope called OW, 'one of the best academic bargains in the country.'"
- University of St. Andrews is hot because "More than a third of the students at St. Andrews come from abroad, and one academic year's fees total less than $25,000."
- Washington and Lee is hot because "funds went to establishing the merit-based Johnson Scholarships, which promise a full ride to about one-tenth of freshmen each year."

- University of Georgia is hot because "$2.9 billion in aid has been meted out to students in the past 15 years."

Clearly hotness has something to do with price, but that doesn't explain why no state university is considered hot in comparison to any private university. State universities are far cheaper and often quite good. And that certainly wouldn't explain why Brown was hotter than Yale every now and again.

I found "being able to do undergraduate research" to be the funniest explanation of hotness. Why is that important, exactly? And if it is important to a student, wouldn't that be the kind of student who ought to attend a research university?

College counselors, the media, and the general paranoia about college that runs through high schools these days, have made college selection a complex and frightening business. So here, without regard to "hotness," I will make a few points about how to choose a college.

1. Don't put yourself into debt to go to college. Price does matter. If you can't afford Yale, don't attend Yale.
2. Know what a college actually offers. Attend a research university because you think you might want to do research later in life. A list of the top 50 research universities can be found in U.S. News and World Report. They are mostly the extended Ivies and the important state universities. If you aren't interested in research, go somewhere else. I know Yale is a nice brand name. If you want a nice brand name, go there, but there are plenty of other places that will educate you just as well.
3. Know what you want to be educated in. Do not go to college with no idea of what you want to learn about or what you are interested in doing later on. If you do that you will major in "sex and drugs and rock and roll" like everyone else, and you will waste your time and your parents' money. You can always put off college until you do know what you are interested in learning.
4. When you think you know what you want to learn, find out if the people who are good at it actually teach at the place you want to go. People say that schools are "good schools" without having a clue what the criteria might be. What is good for you may not be good for the next guy. You must know what the school is good at teaching. Find out.
5. Choose a place that looks like you. Visit. See what the students look like. They differ from place to place for many reasons. Find out where you feel comfortable.

Do not go to college because everyone you know is going to college. Go with a purpose. And – avoid "hot" schools.

The Big Lie: teaching never matters in university rankings

There was an almost perfect article in the *BBC News* online, titled "What Makes a Global Top 10 University."[1] It starts with the usual list of top universities:

> The Massachusetts Institute of Technology (MIT) is in first place in the latest league table of the world's best universities. It's the third year in a row that the U.S. university, famous for its science and technology research, has been top of the QS World University Rankings. Another science-based university, Imperial College London, is in joint second place along with Cambridge University. Behind these in fourth place is Harvard University, the world's wealthiest university. And two more U.K. universities share joint fifth place, University College London and Oxford.

Then, it goes on to list how these rankings are calculated:

> But how does a university get to the top of the rankings? And why does such a small group of institutions seem to have an iron grip on the top places? The biggest single factor in the QS rankings is academic reputation. This is calculated by surveying more than 60,000 academics around the world about their opinion on the merits of institutions other than their own. Ben Sowter, managing director of the QS, says this means that universities with an established name and a strong brand are likely to do better. The next biggest factor – "citations per faculty" – looks at the strength of research in universities, calculated in terms of the number of times research work is cited by other researchers. As a template for success, it means that the winners are likely to be large, prestigious, research-intensive universities, with strong science departments and lots of international collaborations.

So, just to be clear, the rankings are all about how much research money the professors have to play with and how famous they get by utilizing that money. I was a professor at top-ranked universities for 35 years. I brought in lots of money. I was cited often. And I helped my university in its rankings. Great. What about teaching students? Never mentioned. I don't mean never mentioned by this article, I mean never in my 35-year career did anyone mention it. In fact, the more money you bring in and the more famous you are, the less you have to teach. That is one reason poorly paid adjuncts are teaching more these days.

I like this article because it says just this quite plainly:

> Is that a fair way to rank universities? It makes no reference to the quality of teaching or the abilities of students?

1 http://www.bbc.com/news/business-29086590

The idea is that a student wanting to find an undergraduate arts course isn't really going to learn much from rankings driven by international science research projects. Then I saw this:

> Those that focus on teaching rather than research will not be as recognized.

They leave out what I like to call "the big lie of college education." No one ever tells a student applying to Harvard that he will be taught almost exclusively to do research because that is all his professors do or know how to do. He will not be taught to program by his computer science professors, because they haven't written a program in years. He will not be taught how to start a business by his business professors, because they have never started one. He will not be taught anything about being a medical doctor, because medical doctors don't teach undergraduates. In fact, he will not be taught anything of any use in his life at all in college, unless he plans to become a researcher.

I never planned to become a researcher, but that's what I learned in college so that's what I became.

So parents, when you decide to fork out $50,000 a year for Harvard, remember that Harvard's goal will be the creation of a researcher. Do yourself and your kid a favor and ask him if that's what he wants to be when he grows up.

Last week I had a business meeting in New York and, as I often do, I asked the people I was meeting with where they went to college. One said Harvard. He had majored in Economics. So I said, "You got to meet some Nobel Prize winner and hear all about microeconomic theory. Do you use any of what you learned there today in your work?"

"No" he said.

Isn't it time we started talking about teaching? And while we are at it, could we make that conversation be about teaching things that you might actually use as an adult?

Stop cheating undergraduates of a useful education

Undergraduates are being cheated. As a professor of 35 years, I always knew that I didn't care much about undergraduates. I hoped maybe someone else did. The reason I didn't care about them is pretty much the same reason that most professors at research universities don't care about them. It's not our/their job. No, really. It isn't.

I realize that people who have gone to college and then moved on will think that what I am saying is crazy, but they don't really get how a research university works. Professors at top universities run large research labs (in the sciences anyhow). They spend a great deal of time raising money for these labs, keeping sponsors happy, and then actually running the lab and managing all the graduate students and researchers that their money pays for.

Usually, these professors are asked to teach an undergraduate class from time to time. It is the last thing on their minds to worry about, and most would admit they don't do the job well. They know that lecturing and then testing students to see if they listened is not really education. Education is what happens when they help their PhD students individually with their research problems.

I am writing this while visiting an excellent research university, and after just having been at another excellent one. The only time undergraduate teaching comes up in conversations I have with faculty is when I bring it up, and there is simply a collective sigh. How do you teach 200 students sitting in one room?

Universities like the tuition these kids pay, and fool themselves into thinking that they are prepared for something after taking 40 random courses, even though they are certainly recommending graduate school to these students as their real option.

The other day, an official at a lesser-ranked university asked me in passing what I would do if I were to create a computer science department at his university. It was an odd question, to say the least. I have only been a professor at top research universities and the answer there would certainly be to go after some great researchers and start building a great graduate program that was well-funded by outside money. But there are plenty of such departments, and these days my mind goes to education rather than research.

So I proposed something radical. I suggested we could build a computer science undergraduate program that taught students to program. Currently, students sign up to be CS majors because they like programming, and then they are led into theoretical courses or arcane research courses by the faculty. (All in the supposed interest of breadth and readiness for some imagined future.) I suggested that instead we combine the student's interest in building stuff with working people who need stuff built. This particular school happens to have a great medical school, for example. There are lots of opportunities to build important medical software and medical apps. This would happen by letting the CS undergraduates hear about the issues in medicine these days and helping them interact with medical students and practitioners. There is also a great business school at this place. Students could learn how to invent new software for use in business and also how to fund a company to market what they build, possibly by partnering with students in these other schools.

So my idea boils down to this: Let undergraduates do what they came to college to do. In computer science this is actually rather simple. They love computers in the first place, so let professors simply help students do interesting things that make them able to build a product and run their own company, or else become valuable employees in a company that might employ them.

Harvard won't go for this. Where are the liberal arts? What about discussing great ideas? Fine, go to Harvard for that. But it is time that some universities start paying attention to undergraduates in exactly this way – by helping them be what they want to be and attain skills and practical experience in what interests them.

How academic research has ruined our education system

I gave a speech in Barcelona last week. (Actually I gave a lot of speeches in Barcelona last week.) In one that was meant for the general public (at least, I thought that was who was in the audience), I was asked a question about what was new in what I had said. I was speaking about the new experiential MBA program we are developing for La Salle University in Barcelona, which will soon be available as an online web-mentored degree program. And I was talking about the 16 cognitive processes that need to be mastered in order to think, and how they are used in our new MBA.

A member of the audience, who identified himself as a professor at a local university, asked me "what is new here?" The question took me aback. Last I heard, the world was offering classrooms, lectures, tests, and courses as a way of teaching people, not degree programs that were entirely online and project-based, in a story-centered curriculum. I had no idea how to respond. The question was beyond absurd.

My colleague, Sebastian Barajas, who had presented before me and speaks the local language, jumped in and answered the question. I was relying on simultaneous translation, and got most of what was said, but not all. Whatever he responded, however, was not sufficient to quiet this man, and he came back again with more or less the same question. I responded without mercy. Needless to say, impolite behavior is not well-loved in Europe. No one who knows me has ever thought of me as being especially polite, but my response was over the top, even for me.

A flurry of tweets and blog chatter (in Spanish) later erupted about this.[2] The summary of this discussion, for those who don't want to subject themselves to the awfulness of "Google Translate," is that some people like me and some don't – no surprise there. But, what I learned from this was that the audience was made up of academics, and not the general public, as I had thought.

I wondered about my own overreaction. I attributed it to the fact that I was quite ill at the time, having eaten something bad the day before. But after reading the blog I realized what the real issue was, and felt the need to write this rather long column.

This is what infuriated me. The question he asked is exactly the question that has killed off our schools. Now it is a long road to my point here, so please bear with me.

2 http://trabajocolaborativoenred.wordpress.com/2010/02/06/no-todo-el-mundo-quiere-a-roger-shank/

Jonathan Cole, the former provost of Columbia University, and a man whom I like and admire, recently wrote a book about research universities in the U.S., called *The Great American University*.

I haven't read the book, just the review, but I am hardly surprised by, nor do I disagree with, the idea that the research universities in the U.S. are a great treasure that we should endeavor to take care of.[3] But having said that, it is the fault of the research universities that our education system is so awful. And it is the question, "What is new here?" that is the essence of the problem.

Here is what occurs in research universities. Students write PhD theses that are supposed to be original research in their field of choice. I have supervised over 50 PhD theses myself and have read and advised on many more. I am familiar with the process and understand what is wrong with it.

Here's the problem. Think about how many PhD theses are produced each year around the world. Thousands? Tens of thousands? How many of these can really be original (much less be important)? Original is defined as something no one has ever written before in an area of research. So there are many PhD theses, particularly in Europe, that are simply examinations of what other people have said over the years, together with a 'new' point of view. Even in the best science universities, PhD theses are often about minutiae so incomprehensible that no one, not even the graduate students who wrote them, will ever look at them again. There are interesting PhD theses of course, but they are few and far between. The old joke is that a PhD thesis is a paper written by the supervising professor under very difficult conditions.

Why do I mention this? Because it turns out that the man who had asked the "what is new here" question had just written a PhD thesis in which he had examined various education theories, and his conclusion was that because my work is derivative of Plato and Socrates and Dewey – something I am quite proud of, by the way – it wasn't presenting any new theory, and therefore was not worthy of serious consideration. This is kind of like asking, in a disdainful way, at an unveiling of the new 787, "What is new in the theory of flight here?" Only a newly-minted PhD would even think such a thing. Real-world applications and innovative changes matter a lot.

But in the academic world, the questions are always about theories, and applications are looked down upon, which is odd, because no one asks Google "what is new here" (the answer is "not much, since the 1950s"), but it sure has changed our lives.

Research universities produce PhD students who do original work. This would be fine and dandy if there were only a few research universities and if teaching was not affected by this process. But, while there are maybe 25 serious research universities in the U.S., there are hundreds that think they are research universities, and because of that there

3 New York Times Review of "The Great American University" http://www.nytimes.com/2010/02/07/books/review/Goldin-t.html

is pressure for the faculty to do research and publish research. That pressure and that focus are the cause of the awful education system we have in place today around the world. Let me explain.

I will tell you about a friend of mine named Bill Purves. I met Bill when he came to work as a post-doc at my Artificial Intelligence Lab at Yale. He was a professor of Biology at the time and has since retired. People apply to work in prestigious labs all the time, but it is odd to find a biologist who wants to study AI. I was intrigued so I agreed to host him. Bill had a PhD from Yale and I thought he might just be feeling nostalgic for his old haunts. He had been a professor at what you might call secondary research universities. There are 3,000 colleges in the U.S., and at least the top few hundred are trying very hard to act like Harvard and Yale. They hire professors with PhDs from Harvard and Yale who then teach their students what they learned at those places. To put this another way, while the students who go to a state university expect to learn job skills, they are being taught research skills and theories, which are all that their research-university-trained professors actually know.

Bill's research area had to do with some arcane feature of cucumbers. He had been doing work on cucumbers for over 30 years at that point. He wasn't coming to visit me to learn more about cucumbers.

Bill had recently moved to a more teaching-oriented college and had begun to worry about how people learn. We worked on learning at our lab, so he came to learn what we knew. But it was pretty unusual to encounter a research professor who was so worried about teaching. They are more typically concerned with teaching their PhD students (via the apprenticeship method) than they are with lecturing to undergraduates. They pay their dues by teaching so that they can do their real job – which is research. This is a very common attitude.

I am not criticizing here. That was how I felt as a research professor. My job entailed some teaching but was not really teaching. Professors at research universities think about their real work and teach about their real work. This is fine if they are training future researchers. They can criticize their students and each other about whether new research is good work, and they can compete to get their papers published and their presentations at conferences taken seriously. But really, do you think that Bill's biology students cared about cucumbers?

This is a real problem. Students are taught about what interests the professor, but what interests the professor bears little relationship to what students came to college to learn. Perhaps they want to be psychological counselors, but they will be taught about experimental research methods. They may want to become writers but they will be taught about literary theory. They may want to become doctors but they will have to learn about cucumbers, at least they will if a cucumber specialist is teaching the biology course.

Now Bill was not like this at all. In fact he came to Northwestern when I moved there for another post-doc year, to learn more about what we were doing with computer courses. Eventually he became the lead designer in our high school online health-sciences course.

But Bill is a rare bird. He fundamentally cares about students. I am not saying that professors don't care about students. I am simply saying they care about their research more. If students attend a research university they should know the truth. But Yale and Harvard don't really explain the research orientation in a way that would help incoming undergraduates grasp its significance.

I am not concerned about Ivy League students. They get a good education any way you look at it. But if every professor in every major university is playing the same game, then students at, for example, the University of Illinois, ought to know that every one of their professors cares more about research than they do about teaching, and that the "what is new here" standard of assessing work will emphasize theory over practice every time. Since colleges hire professors trained for research, every college is dominated by an issue that detracts from good teaching. Students want job skills and professors teach research skills.

This is why I was furious at the "what is new here" question. Two-thousand years ago, Petronius asked why Roman schools taught so little of what would be useful in everyday life. Nothing has changed, precisely because it is intellectuals, and not practitioners, who dominate the teaching landscape. Students are forced to learn things they have no interest in, in order to get college degrees.

In the end that is why we built a new MBA program. Even MBA programs emphasize theory over practice! This may sound hard to believe, but professors in a business school also tend to have PhDs from Harvard and Yale, and may never have actually run a business themselves. This is why I was asked to design a new learn-by-doing MBA program for La Salle. It is a program designed to upset the status quo. No courses, no theories, simply learning how to do what business people do.

What is new here? If this had already been done, we wouldn't have done it. It is new in the same way that the 787 is new. It is useful and important because it changes the way education works, not because it presents a new theory.

We must fight against the university professors who wish to dominate the discussion of what education should be like. Research professors should be encouraged to do research, but do we really need so many of them? Twenty-five research universities is a fine number for the U.S., and Spain should have maybe one. But if every professor feels obliged to publish more about his work on cucumbers, the world loses its best teachers. Bill Purves understands that and he devoted the end of his career to teaching – not to cucumbers.

Humanities are overrated

Here is a part of an article from *The Chronicle of Higher Education* that came out today:[4]

> The results of an important new cross-disciplinary survey of humanities departments make it clear that the humanities remain popular with students and central to the core mission of many institutions.
>
> ... the bad news ... The survey also found less-than-rosy job prospects for the rising generation of scholars.
>
> ... the good news: The great majority of the humanities departments surveyed—87 percent—said that their discipline was included in the core requirements at their college or university.

I would find this article hilarious if it weren't so sad. But it is a very good example of what is wrong with our university system. There are no jobs for English and history majors and no faculty openings for PhDs in those fields but, nevertheless, the humanities survive at universities. How do they survive? By making the humanities offer required courses that every student must take.

There is nothing wrong with the humanities in principle. We imagine that people might learn more about life, learn to be better people, learn to understand issues that have plagued mankind, and think about what it means to be human. So the humanities must be good stuff, right? Here are some courses picked at random from the Yale catalogue:

ENGL 265b, The Victorian Novel
ENGL 158b, Readings in Middle English: Language and Symbolic Power
ENGL 305b, Austen & Brontë in the World
ENGL 336b, The Opera Libretto
HIST 166Ja, Asian American Women and Gender, 1830 to the Present
HIST 168Ja, Quebec and Canada from 1791 to the Present
HIST 201Ja, The Spartan Hegemony, 404-362 B.C.
HIST 202Ja, Numismatics

I am sure that these are fine courses taught by serious scholars. But that is exactly my point. When people glorify the study of the humanities, they fail to mention that these are scholarly subjects that are of very little use to the average college student. Universities require that students take them because universities don't want to fire the professors they already have, and those professors need to teach something. But, with a few exceptions, they are not teaching students to think constructively about life. Rather,

4 Humanities Remain Popular Among Students Even as Tenure-Track Jobs Diminish http://chronicle.com/article/Humanities-Remain-Popular-A/64419/

they are teaching students about a narrow part of the scholarly domain in which they do research.

Here again we have the clash between the research university and what students expect to learn when they go to college.

The Chronicle of Higher Education represents professors, and they think it's great news that students are being required to take the courses professors want to teach. I think this is awful news. Students need to learn to live in the real world. There are very few scholarly jobs, so there is no practical reason to teach such courses. If these courses taught human skills, as we all assume they do, that would be great, but they don't.

Scholars need to stop running universities.

As I have said many times, I don't think Yale has to change. We do need to produce some scholars, after all. But there are 3,000 colleges in the U.S. that are copying Yale's model. This is too many.

"What should I go to school for?"

These days one can easily find out how people get to one's website. My outrage column is often found via the question, "What should I go to school for?" This question drives the answer seeker to my column on "Why little girls shouldn't go to school," which is certainly not what they were looking for. (I don't think little boys should go to school either, in case you were wondering.)

So, I thought I would attempt to answer this question, since people keep asking it. The problem is that the question is ambiguous. They could be asking why go to school at all, or they could be asking what they should study in school. As I have no idea which meaning predominates, I will take a shot at answering both questions. I will assume that the people asking these questions are in high school and perhaps thinking about going to college.

Why go to school at all?
In a society other than ours, this is a very good question. I think school, as it exists today, is a very bad idea. Still, I would be remiss in answering this question by saying "drop out." Dropouts are viewed badly in our society. School is stupid, but dropping out is stupider. Why? Because as one travels through life, one accumulates a set of accomplishments. Quitting, no matter what you quit, is never a great accomplishment. Unless, of course, you quit for something better. If you have a good plan that will get you something better and enable you to say "I quit to start Microsoft" or the equivalent, by all means quit. One learns very little of value in high school, but the credential entitles you to a minimal amount of respect that you may need at some point. So stick it out if you can.

What should you study?
What should you study in high school or, more importantly, because there are more choices, in college? Let's start with what you shouldn't study. Study no academic subject. Do not study English, history, math, physics, biology, or any of the other standard subjects that one always starts with in high school. Whoa! Did I really say that? Heresy. So, why not?

It is important to realize that there are many myths in our society, and that these myths are usually offered by people who stand to gain from them. The "you must drink eight glasses of water each day" myth, for example, is offered up by companies that sell bottled water. In school, the significance of studying literature, math, history, or science is offered up by those who teach these subjects, those who make money testing these subjects, and book publishers and others who have vested interests in selling things related to these subjects. In addition, the educated elite, having been educated in these subjects, can pooh-pooh anyone who doesn't know them and keep the high ground for themselves. If you don't know what they know, you can't be much. This attitude has always been with us, in every society, but the subjects change. Sometimes the subject is religion, sometimes astrology, sometimes secrets that only the village elders have. These days it is literature, which certainly won't last; mathematics, which makes hardly any sense at all in the age of computers; and history, which never made any sense, since history is written by those who come out best in its telling . Science seems to be making a big move these days. When I was young, science was for geeks, and those involved in it were looked down upon by the people who knew important stuff. Things change.

There is, unsurprisingly, a serious lack of employment possibilities in those areas of study. So many people have been pushed to study those subjects that there is a serious oversupply of job seekers who were English majors, for example. It should not be possible to be an English major, but try telling that to English professors.

So what should you go to school for?

This is really an easy question to answer. First, ask yourself what you really like to do in life, what you think about on a regular basis, who you admire, and who you wish to be. Only you can answer those questions. When you come up with answers, ask if there are jobs in that area. Be creative. Make up a job if you don't think one exists. Ask what you need to learn in order to become a person who thinks about and does whatever it is you like to think about and do. Extrapolate up. If you like working on your car, maybe you would like working on airplanes or ships, for example. If you like hanging out and talking, ask yourself who gets paid to do that (salesmen?). Find out where those who do what seems fun learned to do it. Often the answer is "on the job." If that is the answer, ask yourself how you can get a low-level job in that area and work your way up. People learn by doing. Ask how you can start doing.

If you do need training to start doing what you want, find a community college that offers that kind of training. Most of all, do not go to school if you have no inkling about what you would like to learn. Work for a while and start finding out more about the world, then ask the above questions again.

In the U.S. most people go to college immediately after high school. My experience as a professor was that those students who did something else, who went into the army, the Peace Corps, traveled around, worked for a while and such, made much better students in college. They knew why they were there. Do not go to school if the only reason you are there is to get a degree. Wrong reason. Know yourself first, then learn what you need to know to become a person who you would respect.

Engaged learning

I think I have been barking up the wrong tree. Here I am trying to fix education when I suddenly realize that all you need is a good marketing campaign. Why make real change when you can just say you have?

It was tonight's presidential debate, sponsored by Cleveland State University, that taught me this. Behind the speakers there was a sign that said "Cleveland State University: Engaged Learning." I noticed it because NBC has been using it as a backdrop over the last few days.

I know nothing about Cleveland State, but I am quite sure that it has boring lectures, absurd requirements, many professors who don't care, and students who are just looking to get through the system by jumping through whatever hoops are put in front of them, as is the case at every university I have ever known.

So I wondered if they actually did anything different at dear old CSU, and I went to their website to find out. This is what engaged learning is:

> At CSU, "Engaged Learning" means that whether you are a student, faculty member or staff, you can expect to be an active participant in your learning experience. You can expect to engage in ways that will differentiate your experience at CSU from older, larger, and less diverse learning institutions. You can expect your learning experience at CSU to be distinctive.

Okay. Not bad. "How?" I wondered.

In four important ways, I learned from the university website:

1. An engaged learning logo will be on all communications materials. CSU will unveil a new advertising campaign this spring.
2. The $200 million-plus master plan is remaking the main campus of Cleveland State University.
3. CSU offers more than 140 opportunities to be engaged on campus through a myriad of organizations formed around common interests.
4. A website for engaged learners where they can say what they like about their CSU experience.

And that's it, folks. No new kinds of courses. No new kinds of experiences that eliminate courses and tests. No rethinking of what college should be and what students need to learn how to do. No change of any actual kind. Just money spent on advertising and buildings. Of course, this is real change from my experience. Yale, Stanford, and Northwestern don't advertise (except for revenue producing programs).

But the marketing phrase is so nice: engaged learning! I wonder how much it cost?

Confused about what college is about? So are colleges.

This week we have had a fascinating set of stories emanating from two major U.S. universities. These stories make clear why our conceptions of college are muddled. Since many of my readers do not live in the U.S., I will briefly summarize these stories.

1. Brigham Young University, a university run by and for the Church of Jesus Christ of Latter Day Saints, suspended one of its star basketball players (on a team headed for the national championship) because he had sex with his girlfriend.
2. Mike Bailey, a psychologist at Northwestern, had a live sex demonstration in his class on Human Sexuality.

How are these stories related? There has been much discussion of them, not necessarily in the same articles, but as they happened at the same time there have been some comparisons being made in various publications.

My connection to these stories is not too great, but as I was a member of Northwestern's psychology department, I am familiar with the Northwestern scene and with Mike Bailey. And, although it hardly makes me an expert on BYU, I did spend a few days there recently, interacting with faculty and administrators and generally discussing education.

BYU has a strict honor code to which all students must adhere that stems from their church's religious beliefs.

Northwestern is a more typical U.S. campus, which means students come from everywhere and from every culture, and all live together and interact with each other in the way that kids in their late teens and early 20s who have no real supervision are likely to act.

As a professor, I always felt that kids should be kids, that they should enjoy sex and drugs and football if they like, but that it would be nice if they didn't confuse those activities with getting an education. Alas, there is nothing I can do to persuade people that kids who are on their own for the first time should probably not be going to college. It would be far better if they got the partying out of their system beforehand, and pursued serious education when they were ready to be more serious.

So, I am more in tune with BYU's philosophy than with Northwestern's, only because I think university education is wasted on students who are preoccupied with growing up and finding out who they are (and drinking excessively in the process).

But my viewpoint is actually irrelevant to both of these stories.

The real issue behind these stories is determining the answer to the question "what is college really about?"

At BYU the answer is, one would suppose, preparing students to be productive citizens who live within the rules and philosophies of their particular community.

At Northwestern, the answer would be, one would suppose, the same, except the community is much broader with much more varied rules and options.

But I can tell you, neither of these schools actually does this.

At BYU, when I spoke there, I chided them on copying, more or less verbatim, the curriculum offered at Harvard and Yale. One obvious reason they do this is that their faculty have PhDs from such places, so they teach what they learned there. But the goal of Harvard and Yale is, pretty much, to produce scholars, and possibly to produce future leaders of the country.

BYU exists in a place and in a community that needs a much different approach to education. They are not producing the nation's scholars, and while they may produce some national leaders (Mitt Romney comes to mind), that isn't an everyday occurrence, nor should it be their goal.

I think BYU is right to teach and enforce the rules of its particular world, but curiously, they fail to do this, in that the education they provide is more or less the same as that offered everywhere else.

At Northwestern, the focus should be on producing people who can get jobs that exist in the real world, and making creative people who can function well within that world. Yet Northwestern emphasizes scholarly pursuits, and it offers a smorgasbord of courses that allows students to pick and choose ones that seem like the most fun. Of course, Human Sexuality seems like fun. And since the students actually do need to learn about sex, it makes sense to have such a course.

But that course exists along with thousands of others that are about random topics, which don't fit into a coherent whole that could give students any idea of what they should or can do after they graduate. Northwestern doesn't care that much about producing people who can go to work. They just let the faculty offer the courses that they want to teach.

Mike Bailey has been pushing the envelope on his teaching for some time. He seems to like the ruckus he causes, and personally I don't blame him for actually teaching what he is supposed to teach.

But the fact is, he will be censured in some way for doing this because Northwestern, like most universities, is really about getting students to know things rather than getting students to do things.

The real problem in university education is that no one knows what it is really for any more. It used to be solely about making scholars. Now that the masses go to university in extraordinary numbers, university education is about appealing to the masses. This means providing entertaining courses, and Mike Bailey, while he will likely get into trouble for it, has done just that.

BYU, on the other hand, has actual principles. They are not my principles but why should they be? They are at least trying to do more than entertain. Or, at least, they should be. But they offer the same stuff that Northwestern offers, more or less.

Perhaps it is time to rethink college education, and ask what it is really there for and what students are actually supposed to gain from the experience. When we answer this question, we might want to consider what they will actually do with what they have learned after they graduate.

Just a thought.

Preparing for a fictitious college

I can't tell you how sick I am of hearing about the role of high schools in preparing students for college. Legislators and public officials, who presumably have actually been to college, are worried that students are not prepared for college. This is their excuse for teaching algebra and chemistry and world history – preparation for college.

So, why don't we consider the professor's view on this? When I see a new student in a college class, am I concerned that he may not know the quadratic formula, the Battle of Hastings, or how to balance a chemical equation?

Of course not.

Preparedness for any class I ever taught would mean knowing how to express oneself in an articulate manner, being able to write clearly, being capable of an original thought, being able to reason logically, and the willingness to work hard to accomplish something.

Is that what the high schools send us? No.

Why not?

Because they are busy preparing students for a college that doesn't exist. In this fictitious college, students are left behind because their algebra skills were found wanting and because they had bad SAT scores.

Yes, SATs predict success in college. You want to know why?

The willingness to do the mind-numbing memorization and test practice needed to do well on the SAT does indeed predict the willingness to do whatever your professors tell you to do in college. A personality test would predict this as well.

Now I will tell you a dirty little professor's secret. Professors assume, whenever they teach freshman (this is true of teaching first-year graduate students also) that the students have been poorly prepared, and they start at the beginning.

Please stop confusing getting into college, which has an associated array of silly hoops attached to it, with being prepared for college, which ought not to be the role of high school anyway unless it means learning to think critically and be articulate.

Only Harvard and Yale lawyers on the Supreme Court?

It is not everyday that I feel the need to defend our educational system but I heard something so outrageous on the Today show this morning that I am afraid I must do just that.

A question was discussed between Matt Lauer and Joe Biden on the problem of having a Supreme Court full of Harvard and Yale Law School graduates. They agreed this was "elitist."

I have heard stupid stuff come out of the mouths of politicians and news anchors before, but this one breaks new ground. Maybe in 1920, when Yale and Harvard kept out people who weren't WASPy enough or rich enough for them, such a statement might have made sense. But, while I am the first to criticize our universities for making the entire high school system insane, our problem does not stem from professional training, which is actually where those institutions shine.

Harvard and Yale are creating all these constitutional lawyers because they have a competition and select the best and the brightest from all over the country and, these days, are no longer discriminating against people who were not born rich and white. It is their job to take in the brightest people they can find and produce the best legal minds they can produce.

No one willingly attends the University of Alabama Law School when they could have attended Harvard.

Producing the elite is what these schools are good at. Why shouldn't the Supreme Court be composed of only our smartest and best-educated lawyers? What is Biden thinking?

Please don't make me be a dentist!

I attended a family occasion the other day. I saw people from one side of my family, most of whom I hadn't seen in some years. My first cousin introduced me to her grandson. I was told that he was graduating college and would soon be attending dental school.

I broke out laughing.

Behind him were his two younger brothers. I asked if they would be going to dental school as well. At this point his mother chimed in that she certainly hoped so.

Now I was just sad.

Rest assured, I have nothing against dentists or dental school. A fine career choice I am sure. I have left out some information here. The mother of this boy is a dentist. I also left out that his father is a dentist. I also left out that his grandfather is a dentist. And, I left out that he (and I) have other cousins who are dentists as well. My uncle was dentist. His son is a dentist. His sister married a dentist. Her son is a dentist.

All these dentists are perfectly fine human beings and they all seem to be living well. It is funny to come from a family of dentists – but really, so what?

At some point in the party we were all attending, as the music blasted and people danced, I saw that the young man I had been introduced to had sat down next to me. He said his grandfather had told him that I was some kind of professor and he asked me what I taught. After some chit-chat, I asked him if he really wanted to be a dentist.

He said he had worked hard in college, struggling through required science courses, and that it would soon all be worth it.

I asked him if he had ever considered any other profession. He said no. I asked him why not, and he said there had been a lot of pressure from his family to be a dentist. I asked why and he said they had had good experiences and it had worked for them and they thought it was a great life.

I asked if there was anything else he could imagine being. He replied that he really wanted to work with people and that he liked talking to people, and as he went on I got the idea that it wasn't the teeth part of people that he was referring to.

I told him that when I taught at Yale I devoted one class every term to the subject of what the kids in the class wanted to be when they grew up. I challenged them to be something other than what their parents wanted them to be. But for the most part, the children of doctors were going to be doctors and the children of lawyers were going to be lawyers.

We don't realize as parents how much we talk with children about what they are going to be when they grow up, and how much we limit their choices by talking about the limited things we actually know about, or by inadvertently putting pressure on them to look at the world in a certain way.

When I suggested that this young man not make any choice right now except to decide all this in a few years while trying some other stuff out, he was mostly concerned about how he would explain this to his parents.

Now, I usually write about schooling in this column, and this one is no exception. Except for my weird one-day class, students at Yale got no real career counseling. They only got role models (who were all professional academics), or pressure from their parents, or advice from their peers about what was a hot choice.

Why aren't we teaching our children how to think about career choices, or life choices for that matter? Because we are too busy teaching them calculus or macroeconomics.

Governments complain about the lack of skilled workers but they don't try to help in any way, except to push more math and science courses that are irrelevant and that don't help one understand one's career options. Calculus is not a career choice.

Schools need to start helping kids figure out what they can do in life, or the advisors will all be parents who are limited in their worldview.

Students: Life isn't a multiple choice test. Have some fun.

I want to consider four separate things I happened upon this week that all lead to one glaring conclusion.

The first was this article in the *New York Times*: "More College Freshmen Report Having Felt Depressed"[5]

> High numbers of students are beginning college having felt depressed and overwhelmed during the previous year, according to an annual survey released on Thursday, reinforcing some experts' concern about the emotional health of college freshmen.

This would be very interesting if it weren't so sad. The article reports that students are stressed out about getting into college and academics and so they socialize less and don't even have time to watch TV.

To put this another way, we have managed to test these kids to death, so that their life is all about getting into college by getting good grades. How does this make for well-adjusted human beings? Do families even gather around the dinner table and talk anymore? Do they play together after dinner? Or are they all cramming for the next test? What kinds of people are we raising? If you are depressed when you arrive at college, how are you going to even get through college much less life? Where is the fun?

Well, apparently not in childhood. The next article, also from the *New York Times*, makes it clears why, "Is Your First Grader College Ready?"[6]

> Matriculation is years away for the Class of 2030, but the first graders in Kelli Rigo's class at Johnsonville Elementary School in rural Harnett County, N.C., already have campuses picked out. Three have chosen West Point and one Harvard. In a writing assignment, the children will share their choice and what career they would pursue afterward. The future Harvard applicant wants to be a doctor. She can't wait to get to Cambridge because "my mom never lets me go anywhere."

They are talking about college in first grade? Why? "If you focus on Harvard you will get in," is apparently the answer. But Harvard has a 5.9% acceptance rate. It is probably a lot lower in Harnett County, N.C. So is our goal to get kids to focus on what they will never achieve so that they can be depressed once they get into college, or worse fail to get into college? How can college matter in any way to a six-year-old? Fun matters. Learning

5 http://www.nytimes.com/2015/02/05/us/more-college-freshmen-report-having-felt-depressed.html?smid=tw-share

6 http://www.nytimes.com/2015/02/08/education/edlife/is-your-first-grader-college-ready.html?smid=tw-share

what you want to learn matters. We have made school into a contest that no one can win. Are all Harvard graduates so happy and successful? I don't know. I taught at Yale, where there were a lot of miserable kids and where plenty of the graduates who never went on to do all that much. It is all so sad.

And then I got this, forwarded from my son. It is from his four-year-old's teacher:

> Hello Families,
>
> In honor of Black History Month, throughout the month of February, each classroom at our school will be highlighting important contributions of African Americans to our country and culture. Our classroom will be studying and celebrating the inspiring artwork of Shinique Smith, a Baltimore native who is renowned for her bright, geometric and abstract paintings, collages and sculptures. We are thrilled to introduce Ms. Smith's work to the children as her artistic interests and philosophies are very similar to our students' artistic tendencies in the classroom art studio:
>
> - The children and Ms. Smith share a passion for reusing recyclable materials in their artwork, giving "found treasures" and "loose parts" new life through their creations.
> - The children and Ms. Smith share a fascination with spirals and mandalas, consistently incorporating circular patterns and designs into their work.
> - The children and Ms. Smith have been inspired by the work of Jackson Pollock and enjoy utilizing flicking, splattering and dripping techniques on their canvases when using tempera paint.

I am sure that a four-year-old "studying and celebrating" an artist will be something to behold. But, we have a hint of what will happen. Apparently the four-year-olds have been inspired by Jackson Pollack to dribble paint on canvases. Really? No one just plays with paint any more. Now they are all Jackson Pollack. And since they like playing with junk, we find this is now a tribute to an artist that no one has ever heard of.

This wouldn't bother me so much except for what followed.

> **Extending Learning at Home:**
> Here are some resources to learn more about Shinique Smith at home. Consider taking some time to look through her work with your child (or the whole family). Spark discussion by asking the following: "What does this remind you of?" "What do you see in this piece?" "What do you think Shinique was thinking about when she painted/sculpted this?" "What shades of color do you see?" "What shapes do you see?" "How does this piece make you feel?"

The parents are being told that despite the fact that they have spent the whole day working and despite the fact that the kids have been in school all day, what they should do at home in their free time is this: They shouldn't play with their child or talk to him about what he is thinking, Instead, they should talk about the work of an artist they have never heard of and don't care about to a kid who has no interest in the subject. All this because the teacher wants to rest during class while the kids throw paint?

The good news is that I won't be visiting my grandson or I would talk to him about what the teacher was thinking when she sent home this message or how doing this required art work makes him feel. Fun? That's out. Let's make them stress about school 24/7.

But it hasn't been a bad year for kids in school this year. Why? Because there have been lots of snow days. Kids celebrate when school is cancelled. I wonder why.

But, apparently not in Indiana:[7]

> Even when schools are closed for snow, students in Delphi, Ind., are expected to log on to their classes from home.
>
> The seniors in Brian Tonsoni's economics class at Delphi Community High School are no strangers to technology — everybody has an Internet-connected laptop or smartphone in front of them in class as they work on business plans.

Can we please stop and think about what we are doing to our children? They are all in a giant competition but I am not sure for what. I didn't pay any attention to that competition when I was a student. I graduated #322 in a class of 678. Those numbers never left my mind. I had a C average in college.

Why? Because I believe in playing and having fun and not in stressing out about school. Still I managed to be the youngest full professor at Yale at the time (at 29).

Let them have fun, please. School just isn't that important. I never got into Harvard. (Nor did I apply.) Somehow I managed through life without it. College has become a symbol of achievement in this country. It isn't. There are 4,000 colleges. Anyone can get into college. And anyone can graduate by memorizing answers and passing tests.

Life isn't actually a multiple choice test.

7 For Some Schools, Learning Doesn't Stop On Snow Days http://www.npr.org/blogs/alltechconsidered/2015/02/02/382701005/for-some-schools-learning-doesnt-stop-on-snow-days

How do you know if you are college-ready?

Someone I follow on Twitter (@sisyphus38) (I don't follow many people, but this guy is a frustrated teacher and I appreciate his pain) tweeted today: "Can someone explain what "college-ready" means?"

Here is my response. You are college-ready:

- If you know how to study for, and can pass a multiple choice test.
- If you can sit through a boring lecture and stifle the impulse to jump out the window.
- If you can drink heavily all night and still get up in time for an 8 AM class.
- If you know how to skim a textbook.
- If you have a lot of money in the bank.
- If you know how to go and talk to a teacher to explain why you should have gotten an A and instead of a B.
- If you can figure out how not to show up for class and still pass it.
- If you don't mind not actually doing anything in school except reading and listening.
- If you understand that college is actually pretty much a four-year vacation from the real world and you won't have to do much.
- If you understand that college won't get you a job and there is no point thinking it will.
- If you're capable of doing your own laundry.
- If you are capable of understanding that the university you are attending is not particular interested in you and your needs and you just have to go along with whatever obstacles they put in your way.
- If you understand that the university is mostly about the needs of faculty.
- If you know never to ask the question "Why do I need to know this?"
- If you know that the question "Are we responsible for this?" is the best way to irritate a professor.
- If you know that a professor's opinion will never matter to you ever again.

Or, you could just worry about these two:

- You are college-ready if you have some idea what you might want to do with your life.
- You are college-ready if you know that somewhere in the university there is someone who is willing to take you under their wing and that your real job is to find that person and seek their guidance.

Parents, relax. Your kid will get into college; the question is whether or not to go.

The other day I had dinner with a married couple who were worried about their son. I have these conversations quite frequently although not always at dinner. They were worried about the fact that their son didn't always do what he was supposed to do at school and wasn't getting great grades. They wanted to know if they should pull him out of his fancy private school and put him in public school hoping that he would get A's there.

This sounded like an odd idea. I wondered why they wanted to do that, knowing full well the reason would be that they were worried that he wouldn't get into college. This conversation must go on all the time in homes across the U.S. Everyone has been sold the idea that their kids must go to college. If a kid is bored with school and finds better things to occupy his time, parents panic.

But, this is about to end.

Today, there was yet another article about why, in *The Chronicle of Higher Education*, which is not exactly a radical rag.[8] The article is talking about one particular college and in it there is this:

> The college's administrators say that to achieve long-term financial stability, it needs to expand its enrollment, attracting more students even as competition from other colleges and universities increases. It's a challenge many of the smallest liberal-arts colleges face.

To put this another way, with 4,000 colleges in the U.S. – many of them charging high tuition – anyone who can afford it will find that their kid can get into college.

The real question is whether they should go.

My daughter received a letter from someone with a business that related to hers. It was signed with the name of the woman who wrote it and followed by "Harvard class of 2016." In other words the writer is a junior at Harvard.

Students in college today are not becoming history or philosophy majors in great numbers. We live in a society that now heavily values entrepreneurship and technical skills. You don't need to go to college to do either of them. My company (XTOL) has

8 Survival at Stake http://chronicle.com/article/With-Survival-at-Stake-Small/190491/?key=G29zlFg-6Zn9EY3A2YG1GbjhUYXVkZk57NylfPnogblFXEw%3D%3D==

launched certificate programs under the aegis of well-known universities which are actually teaching real skills and where having a college degree isn't necessarily required. This will become more and more common soon. Students will be able to learn what they want to learn without having to obsess in high school about AP tests and SATs. They will not have to go to college, and if they do go they should have a very good reason to go there, learn what they want to learn, and leave. (Bill Gates and Mark Zuckerberg did exactly that.)

When I was the Chairman of Computer Science at Yale, one year each Chairman had to address the Freshman about why they should major in their subject. I gave a short speech. I said: "Major in Computer Science. Get a job."

I was booed. No student at Yale in 1981 was concerned with such practicalities. That year most of our graduates went to work at this startup called Microsoft.

Am I recommending that kids shouldn't go to college? NO. I am simply saying they (and their parents shouldn't be worrying about this. Colleges will be dying to have them. And, because of that, kids will have the right to blow off high school, which I certainly encourage.

Chapter 6: Death to the Standardistas!

Death to the Standardistas!

The more things change the more they stay the same. Or more accurately, the more we try to change things, the more people who misunderstand the problems in education try to keep things the same. Today's case in point: entrepreneurship education.

Some months ago, we gathered some top business and academic people to meet for a day and do a high-level design for a one-year, story-centered, learn-by-doing, online curriculum in entrepreneurship for high school. We heard about a foundation that was interested in funding entrepreneurship education so we sent them our design, which was more or less an attempt to start a business in a Second Life kind of virtual world and compete with other students in that world. Before that part, the students took on certain business analysis projects to get them ready, and in the final part they tried a real project on the web.

We were told by the foundation that looked at our design that we had not paid attention to the National Content Standards for Entrepreneurship Education!

Of course, we hadn't. We had never heard of them. So, it was with some trepidation that I went online to take a look. I say it this way because standardistas are always wrong-headed and evil. Why? Let me count the ways in which standards are a disaster.

1. They tell you what you must teach and therefore allow no possibility of doing things differently.
2. They are always testing-oriented.
3. They always say what the student must understand and must know and must be able to explain which is a code for we will tell him this and then he will tell it back to us.
4. They invariably do not allow for freedom on the part of the student to get interested in one thing while not being interested in another.
5. They are made by a committee that always insists on listing all the things any person in that field must know without realizing that knowledge comes after doing, not before.

So, knowing my prejudices, now let me show you what I found. Here are some the skills listed that every student must have:

- Explain the need for entrepreneurial discovery
- Discuss entrepreneurial discovery processes
- Use external resources to supplement entrepreneur's expertise
- Explain the need for continuation planning

- Value diversity
- Conduct self-assessment to determine entrepreneurial potential
- Maintain positive attitude
- Explain the concept of human resource management
- Explain the nature and scope of operations management
- Explain the nature of effective communications
- Address people properly
- Interpret business policies to customers/clients
- Use basic computer terminology
- Compress or alter files
- Determine file organization
- Explain the law of diminishing returns
- Describe types of market structures
- Read and interpret a pay stub
- Explain legal responsibilities of financial institutions
- Explain the rights of workers
- Describe use of credit bureaus
- Develop job descriptions
- Encourage team building
- Describe the elements of the promotional mix

There were well over one hundred of these. And, they would, if actually paid attention to, get my nomination for the most boring curriculum ever invented in a fundamentally learn-by-doing field. Further there would be lots of tests, each one saying explain this and describe that.

The lesson here is simple. As soon as standardistas get a hold of a curriculum it will be turned into the garbage that has always been our school system – one where knowing, explaining, and describing always win out over doing and learning from one's own mistakes. The facts and nothing but the facts.

Too bad. I kind of like this field. Kids would have enjoyed running their own businesses in simulation and in reality.

Note to Mr. Foundation Head: After you blow millions on garbage entrepreneurial education by boring and testing students to death just like we have always done, and don't produce even one more entrepreneur, and don't deter even one more kid from dropping out, don't come crying to me.

What is wrong with trying to raise test scores?

I had the opportunity to meet with the great test score promoters in Washington but turned it down. What would be the point? The entire Obama administration is devoted to raising test scores. I would convince no political person to change his or her view.

That said, what is really wrong with testing and emphasizing test scores? Here are five short answers.

Testing teaches that there are right answers
The problem is that in real life, the important questions don't have answers that are clearly right or wrong. "Knowing the answer" has made school into Jeopardy. It is nice to win a game show, but important decisions are made through argumentation and force of reason not knowing the right answer.

Testing teaches that some subjects are more important than others
The tests are small in number. If there were thousands to choose from, then perhaps people could get tested in fiber optics instead of history. But the system has determined which subjects are the most important. Just remember that the system made that determination in 1892. Some things have changed in the world since then. No one in Washington seems to have noticed.

Testing focuses teachers on winning not teaching
Many teachers are extremely frustrated by the system they have found themselves to be a part of. They cannot afford to spend time teaching a student or getting a concept across if the issues being taught are not on the tests. They are judged on the basis of test scores. So, any rational teacher gives up teaching and becomes a kind of test preparation coach.

Students learn that memorization is more important than thinking
Teaching students to reason ought to be the beginning and end of what education is about. But in an answer-obsessed world, "go figure it out for yourself" or "go try it and see what happens" are replaced by more memorization. Giving kids a chance to fail helps them learn. Actively preventing failure by telling the right answer just helps kids pass tests.

Innovation in education is eliminated

How can we offer new curricula and new ways of learning if no matter what we do children must pass algebra tests? The administration says science is important over and over again but since science in high school is defined by boring tests of vocabulary terms and definitions for the most part, who would be excited to learn science? If a really good scientific reasoning curriculum were created the schools could not offer it unless it helped kids pass the very same tests that curriculum was intended to replace.

Oh, one more thing

Testing also reduces knowledge to short answers like the ones I have given here. In reality, serious argumentation is much more lengthy.

Chinese do better on tests than Americans! Oh my God, what will we do?

Recently, we have been subjected to yet another round of fright about our education system because the Chinese have scored better than the U.S. on the PISA test. U.S. Secretary of Education Arne Duncan tweeted "PISA results show that America needs to ... accelerate student learning to remain competitive." The *New York Times* ran its usual scare article. "The results also appeared to reflect the culture of education there, including greater emphasis on teacher training and more time spent on studying rather than extracurricular activities like sports."

I even heard a man who should know better state that these tests were actually meaningful since they were problem-solving tests. Nothing would convince me that the tests were meaningful in any way, but just for fun I took a look at some sample questions anyway. Here are three of them (chosen because they were shorter than the others):

> **Question 48.1**
> For a rock concert a rectangular field of size 100 m by 50 m was reserved for the audience. The concert was completely sold out and the field was full with all the fans standing. Which one of the following is likely to be the best estimate of the total number of people attending the concert?
>
> A. 2,000
> B. 5,000
> C. 20,000
> D. 50,000
> E. 100,000
>
> **Question 7.1**
> The temperature in the Grand Canyon ranges from below 0° C to over 40° C. Although it is a desert area, cracks in the rocks sometimes contain water. How do these temperature changes and the water in rock cracks help to speed up the breakdown of rocks?
>
> A. Freezing water dissolves warm rocks.
> B. Water cements rocks together.
> C. Ice smoothes the surface of rocks.
> D. Freezing water expands in the rock cracks.

Question 7.2

There are many fossils of marine animals, such as clams, fish and corals, in the Limestone A layer of the Grand Canyon. What happened millions of years ago that explains why such fossils are found there?

A. In ancient times, people brought seafood to the area from the ocean.
B. Oceans were once much rougher and sea life washed inland on giant waves.
C. An ocean covered this area at that time and then receded later.
D. Some sea animals once lived on land before migrating to the sea.

Whether or not you know the answers to these questions, it is important to think about what it means to be good or bad at such questions. As someone who studied mathematics and who considers himself a scientist, I can tell you that these questions are both simple and irrelevant to the average human being. One can lead a prosperous and fulfilling life without knowing the answer to any of them. Why then are test makers, newspapers, and Secretaries of Education, hysterical that the Chinese are better at them than their U.S counterparts?

One answer is that every nation needs scientists and that knowing the answer to these questions is on the critical path to becoming a scientist. I can assure you that that is simply false.

Whether or not a nation needs scientists, it surely doesn't need very many of them. In any case, while scientists I know would know the answers to these questions, that has nothing to with the reason they have been successful as scientists. More relevant would be a personality test that sought to find out how creative you were or how receptive you were to new ideas or how willing you were to entertain odd hypotheses. Having been a professor who supervised PhD students from many different countries, I can assure you that Chinese students are very good at learning what the teacher said and telling it back to him. Of course they do well on tests if they come from a culture where that is valued. In the U.S., questioning the teacher is valued and most U.S. scientists have stories about how they fought with their teachers on one occasion or another. If we need scientists, why not find out what characteristics successful scientists actually have? Memorizing answers is probably not one of them. You don't win Nobel Prizes, something the U.S. is still quite good at, by memorizing answers.

But, of course, the problem with the U.S. education system is not in any way our lack of ability to produce scientists. We are very good at it actually.

Our problem is that a large proportion of the population can't reason all that well. We don't teach them to reason after all. What we do is teach them mathematics and science they will never need and then pronounce them to be failures and encourage them, one way or another, to drop out of school. Brilliant. We also to paraphrase President John Adams, don't "teach them how to live or how to make a living."

As usual, neither Arne Duncan nor the new media has a clue about the real issue in education. To paraphrase President Clinton, "It's the curriculum, stupid."

The World Cup of testing

I recently spoke at a meeting in Bogota, Columbia, sponsored by El Servicio Nacional de Aprendizaje[1] (SENA) which is an organization that promotes technical training that leads to useful employment. I had spoken there before and was happy to return since places like SENA are important in the battle against the idea that all students must have an "academic" education.

But, I soon discovered, that Columbia was in an educational existential crisis because they had scored last in the PISA tests. Most Americans have never heard of PISA tests but the rest of the world has. There is a weird competition going on between countries about success at PISA. Here, courtesy of Wikipedia are the 2012 rankings (and scores) in mathematics:

Rank	Country	Score	Rank	Country	Score
1	Shanghai, China	613	19	Australia	504
2	Singapore	573	20	Ireland	501
3	Hong Kong, China	561	21	Slovenia	501
4	Taiwan	560	22	Denmark	500
5	South Korea	554	23	New Zealand	500
6	Macau, China	538	24	Czech Republic	499
7	Japan	536	25	France	495
8	Liechtenstein	535	26	United Kingdom	494
9	Switzerland	531	27	Iceland	493
10	Netherlands	523	28	Latvia	491
11	Estonia	521	29	Luxembourg	490
12	Finland	519	30	Norway	489
13	Canada	518	31	Portugal	487
14	Poland	518	32	Italy	485
15	Belgium	515	33	Spain	484
16	Germany	514	34	Russia	482
17	Vietnam	511	35	Slovakia	482
18	Austria	506	36	United States	481

1 National Apprenticeship Service

Rank	Country	Score	Rank	Country	Score
37	Lithuania	479	52	Malaysia	421
38	Sweden	478	53	Mexico	413
39	Hungary	477	54	Montenegro	410
40	Croatia	471	55	Uruguay	409
41	Israel	466	56	Costa Rica	407
42	Greece	453	57	Albania	394
43	Serbia	449	58	Brazil	391
44	Turkey	448	59	Argentina	388
45	Romania	445	60	Tunisia	388
46	Cyprus	440	61	Jordan	386
47	Bulgaria	439	62	Colombia	376
48	United Arab Emirates	434	63	Qatar	376
49	Kazakhstan	432	64	Indonesia	375
50	Thailand	427	65	Peru	368
51	Chile	423			

My reaction to tests is one of contempt. Why do countries willingly engage in this competition? Why is testing in any way relevant to real education for life? We all assume it is. My reaction to tests as a kid was to simply not care about them. But in those days we weren't terrorizing kids, parents, and teachers about them.

Yesterday an article appeared about Sweden saying their schools were a mess because they had low PISA scores. The acknowledged winner in PISA is Finland which is just a little odd because a check of the scores shows that they are not the winners. I could care less about this game but Colombia was so upset about it that they invited two representatives from Finland to this meeting as well, of course, as well as a representative from China, the country that is actually winning. The World Cup was going on while this meeting was going on, and one couldn't help but notice the analogy between them.

To get a better perspective I have selected one question from the PISA sample tests available online so that we can know what we are talking about here. Here they are, one reading question, one math question, and one science question. PISA scores all three (even though I only showed math rankings above):

Read the text and answer the questions which follow.

IN POOR TASTE
from Arnold Jago
Did you know that in 1996 we spent almost the same amount on chocolate as our Government spent on overseas aid to help the poor? Could there be something wrong with our priorities? What are you going to do about it?
Yes, you.

Arnold Jago, Mildura
Source: The Age Tuesday 1 April 1997

Question: Arnold Jago's aim in the letter is to provoke

A. Guilt
B. Amusement
C. Fear
D. Satisfaction

LICHEN
A result of global warming is that the ice of some glaciers is melting. Twelve years after the ice disappears, tiny plants, called lichen, start to grow on the rocks. Each lichen grows approximately in the shape of a circle. The relationship between the diameter of this circle and the age of the lichen can be approximated with the formula:

$$d = 7.0 \times \sqrt{(t - 12)} \quad \text{for } t \geq 12$$

where d represents the diameter of the lichen in millimetres, and t represents the number of years after the ice has disappeared.

Question: Ann measured the diameter of some lichen and found it was 35 millimetres. How many years ago did the ice disappear at this spot? Show your calculation.

BUSES
Ray's bus is, like most buses, powered by a petrol engine. These buses contribute to environmental pollution. Some cities have trolley buses: they are powered by an electric engine. The voltage needed for such an electric engine is provided by overhead lines (like electric trains). The electricity is supplied by a power station using fossil fuels. Supporters for the use of trolley buses in a city say that these buses don't contribute to environmental pollution.

Question: Are these supporters right? Explain your answer.

My main reaction to these test questions is "I don't care" which would have been my reaction as a kid as well. It is easy to see why countries like Colombia do poorly on them. (Peru is actually last on all three sections in the 2012 Wikipedia article.) The questions are about issues (with a strong environmental bias in every question) that might not be on the mind of your average Peruvian or Colombian. These are countries with populations often tucked away far from the major cities that have difficulty getting any real education out to the provinces. These are also countries whose issues should be more focused on better health, more jobs, and a lot less concerned with preparing kids for a university.

But getting a more realistic focus for schools away from academics and the obsession about college admittance gets more difficult with tests like PISA getting all the schools' attention.

The Finland people at this meeting admitted that it was much easier doing well at the PISA test when you had a very small and homogeneous country. Still, the Colombians wanted desperately to learn from the Finns who really had nothing relevant to tell them.

Testing has become a major industry and Pearson is always lurking ready to make more money on test prep and grading tests, not to mention making them. And, yes, Pearson was at this meeting as well. When I said I would skip what Pearson had to say because they were evil, I got a hearty round of applause.

We really need to stop this testing obsession and get on with letting kids try out things that appeal to them and help them get good at those things. Every student needn't learn the same stuff. Kids have different interests. The best thing we can do for kids is to help them explore what fascinates them. But with PISA lurking we can't let that happen. Losing the World Cup (in testing) is a horrible possibility apparently.

Frank Bruni thinks kids are too coddled. I think kids are too tested.

It seems if you write for the *New York Times* you must write about why Common Core is wonderful. I don't know why. Frank Bruni wrote a column about how today's kids are coddled.[2] I couldn't agree more. Every game ends in a tie. No one can walk anywhere by themselves. Now I am done agreeing with Bruni. Here is some of the nonsense he wrote:

> I behold the pushback against more rigorous education standards in general and the new Common Core curriculum in particular. And it came to mind when Education Secretary Arne Duncan recently got himself into a big mess. Duncan, defending the Common Core at an education conference, identified some of its most impassioned opponents as "white suburban moms" who were suddenly learning that "their child isn't as brilliant as they thought they were, and their school isn't quite as good."

So, this is absurd. Common Core is being fought against because it means school is testing testing testing and what is being tested is boring at best and basically stultifying.

> But if you follow the fevered lamentations over the Common Core, look hard at some of the complaints from parents and teachers, and factor in the modern cult of self-esteem, you can guess what set Duncan off: a concern, wholly justified, that tougher instruction not be rejected simply because it makes children feel inadequate, and that the impulse to coddle kids not eclipse the imperative to challenge them.

More nonsense. People are fighting because mathematics is being rammed down the throats of kids who will never use it. Because science has been reduced to rote memorization and because reading has been made in to a painful activity.

> The Common Core, a laudable set of guidelines that emphasize analytical thinking over rote memorization, has been adopted in more than 40 states. In instances its implementation has been flawed, and its accompanying emphasis on testing certainly warrants debate.

NO. They emphasize memorization and testing. How would like to take tests all day Frank? How would like to learn things that you didn't want to learn just because some testing companies realized that that stuff is easy to test?

---

2 Are Kids Too Coddled? http://www.nytimes.com/2013/11/24/opinion/sunday/bruni-are-kids-too-coddled.html?_r=0

> What's not warranted is the welling hysteria: from right-wing alarmists, who hallucinate a federal takeover of education and the indoctrination of a next generation of government-loving liberals; from left-wing paranoiacs, who imagine some conspiracy to ultimately privatize education and create a new frontier of profits for money-mad plutocrats.

Come on. Common Core is not a right-wing issue any more that it is a left-wing issue. It is a business issue. Bill Gate is behind it and big money is at stake. The idea that kids can learn what interests them is out the window. How is that a political issue?

> Then there's the outcry, equally reflective of the times, from adults who assert that kids aren't enjoying school as much; feel a level of stress that they shouldn't have to; are being judged too narrowly; and doubt their own mettle.

That is a weird idea. Kids should enjoy learning. Of course not. Terrible idea, right Frank?

> Aren't aspects of school supposed to be relatively mirthless? Isn't stress an acceptable byproduct of reaching higher and digging deeper?

No, stress and learning are unrelated. Were you stressed from writing this column Frank? Did you learn anything from writing it? Will you learn anything from what I am writing? Will you find it stressful?

> But before we beat a hasty retreat from potentially crucial education reforms, we need to ask ourselves how much panic is trickling down to kids from their parents and whether we're paying the price of having insulated kids from blows to their egos and from the realization that not everyone's a winner in every activity on every day.

This has nothing to do with the Common Core issues. The curriculum is awful. See if you can pass any of the tests.

> David Coleman, one of the principal architects of the Common Core, told me that he's all for self-esteem, but that rigorous standards "redefine self-esteem as something achieved through hard work."

Achieved through hard work that you want to do, not that you are being made to do. Hard work that accomplished a goal that you have, not that someone else has for you.

> And they'll be ready to compete globally, an ability that too much worry over their egos could hinder. As Tucker observed, "While American parents are pulling their kids out of tests because the results make the kids feel bad, parents in other countries are looking at the results and asking themselves how they can help their children do better."

They will be able to compete globally? In the math competition? We aren't teaching them computer skills or business skills or entrepreneurial skills or invention skills or even social skills. We are teaching them test-taking skills, so maybe they will win the math prize. Hooray!

Measurement in preschool? Measure this!

Testing in preschool? I thought preschool existed so kids could have fun in a safe place while their parents did something other than watch over them. I guess I was wrong.

Last week I had a conversation with my four-year-old granddaughter in which she told me she knew the names of all the planets and proceeded to name most of them. I asked her what a planet was. She had no idea. I asked her older brother. He said they were like big rocks.

Does anyone ever wonder about all this? Must we continue ramming facts down kid's throats so that the people who make tests can get rich?

From the *Washington Post:*[3]

> The (D.C.) board set out to provide parents with a clearer picture of how charter schools compare with one another. It also wants to provide educators with a way to measure progress toward the goal of better preparing children for school, a goal that led city leaders to make a historic investment in universal preschool for 3- and 4-year-olds.

So we will test in preschool now? So more testing companies can get even richer? Last I looked there was $300 million being spent on simply grading tests in Florida alone.

And now preschool? Aren't the testing companies rich enough?

And what is all this preparation they talk about supposed to be for? If you listen carefully it is really about preparing kids for college. At 4?

When I talk about education one-on-one with a professional adult, I often start by saying: I don't know where you went to college or graduate school or what you studied, but let me make a guess: none of what you do every day in your professional or personal life you learned in school. I have never heard anyone respond with anything stronger than, "well maybe a little bit."

Let's be clear about what schools are for. Schools are not for education. We have deluded ourselves so greatly with this myth that we actually think we are measuring something about how well it is working.

---

3 In D.C., public school for 3-year-olds is already the norm https://www.washingtonpost.com/local/education/in-dc-public-school-for-3-year-olds-is-already-the-norm/2013/02/20/e1f84426-7b6a-11e2-82e8-61a46c2cde3d_story.html

You want something to measure? Measure what schools really do:

1. Is my kid being kept safe so I can work (or play)?
2. Is my kid learning to control his impulses and sit still for long periods of time?
3. Is my kid being fed lunch?
4. Is my kid being properly indoctrinated to be a model citizen who can say why the U.S. is a great country?
5. Can my kid defend himself from the bullies?
6. Does my kid have the right clothing so that other kids won't make fun of him?
7. Is my kid being taught enough meaningless stuff so that he doesn't look foolish when asked who George Washington or Abraham Lincoln was?
8. Are they making sure that my kid is really afraid to express an outlandish thought that no one he knows agrees with?
9. Are they making sure that if there is something my kid really wants to do that it will be designated an "after school activity?"
10. Are they making sure that my kid believes that only losers don't go to college?

I suggest we start admitting that these are the real purposes of preschool or any school. Maybe we should start measuring schools on how well they do at teaching them.

"Life is a series of tests." What a load of nonsense.

The *New York Times*, this time in an article by Elisabeth Rosenthal, their former Beijing bureau chief, has waved the pro-testing flag once again.[4] She describes the constant testing of her children when they attended school in China, and notes that while it was stressful, years later they don't recall it as having been awful. Perhaps this was due to the fact they were learning a different culture and language and remember that interesting learning experience more?

Nevertheless she reiterates the *New York Times* party line by saying:

> But let's face it, life is filled with all kinds of tests — some you ace and some you flunk — so at some point you have to get used to it.

I beg to differ.

Life is full of all kinds of situations that test you. Life is not full of multiple choice memorization tests at all.

She quotes experts who argue how testing is killing our children, but somehow, amazingly, decides testing is good. The real question is why the *New York Times* is constantly beating the testing drum. There is lots of money to be made in textbook publishing and testing and those who make big money on that are always in favor of testing and have been the ones pushing No Child Left Behind.

Time to come clean, *New York Times*. How much money are you making on testing?

---

4 Testing, the Chinese Way http://www.nytimes.com/2010/09/12/weekinreview/12rosenthal.html?_r=0

American business hasn't a clue

I saw this press release the other day:[5]

Global Thought Leaders to Convene at MasterCard Math Education Summit

Wow! Who knew there were global thought leaders in math education? I wondered who they were until now.

> The summit features Singaporean math education expert Lianghuo Fan, Ph.D., who will discuss how math teachers help Singapore lead the world in student math achievement. Dr. Fan is an associate professor of mathematics education at the National Institute of Education, which trains all primary and secondary teachers in Singapore.
>
> Other speakers include Frank Corcoran, founding teacher of KIPP Academy in New York City, the highest-performing non-selective public middle school in the Bronx for 10 consecutive years in reading scores, math scores and attendance, and Jason Kamras, the 2005 National Teacher of the Year and special assistant to the chancellor for teacher performance in the District of Columbia Public Schools.

Ah. Global thought leader means someone who makes test scores go up. Got it.

> "There is widespread evidence of the need for improved math education," said Victoria May, director of science outreach at Washington University in St. Louis. "American eighth graders rank fifteenth globally in math achievement, behind students in countries such as Estonia, Hungary and Malaysia."

That evidence is what, exactly? That we are behind in test scores? That is the evidence that we need to win the test score contest!

> "Math achievement is key for students' future employability," said Dr. Terry Adams, superintendent of the Wentzville School District. "Students of today will face global competition for the employment opportunities of tomorrow. This globalization of the workforce means our students must be as proficient in math and science as their international counterparts."

Employment in the future depends on math test-taking ability? Interesting! How do they know this? Are jobs going to those who know the quadratic formula and can balance chemical equations? Why? When did this happen? Or is this just some bizarre notion of the future based on nothing?

> "Technology companies such as MasterCard depend on a future workforce that is skilled in math," said W. Roy Dunbar, president of Global Technology and Operations at MasterCard. " As an innovation and global commerce leader, math is core

5 https://www.mastercard.com/us/company/en/newsroom/pr_FINAL_Math_Education_Summit_Release_10_23_07.html

> to our business. We are not experts in math education, but we recognize how critical the support of math teachers is to improving student math achievement."

Really? Math is core to MasterCard's business? Are they doing a lot of trigonometry over there at MasterCard? Congruent triangles on the MasterCard logo? How exactly is math core to MasterCard's business?

Ah. They mean they like people who can add. I am sure they do.

We really have to stop this math and science nonsense. MasterCard needs people who are courteous, helpful, can work with others, can speak and write, and who understand business. Stop the math and science fixation. Math in America means testing and that usually means algebra. There is no evidence that this has anything to do with America's future. American business needs people who can think, not people who can pass algebra tests.

If MasterCard wants to save the country, maybe they actually look at what their employees do on a daily basis and help students get good at that.

Measure or die

If you want to know why education is bad and getting worse all you have to do is read the questions sent to me by potential investors who were thinking about working with our learn-by-doing experiential model of education.

Now, I have to say that these particular investors are well-known bad guys in the education space and why they were interested in investing in us is anybody's guess. But the questions were so telling, I thought people should see them.

1. What evidence do they have of the effectiveness of their particular programs, if any? What evidence do they have from Socratic Arts of effectiveness of a similar model with a similar student base and similar subject matter areas? How do they plan to gather evidence of effectiveness?

Why do I think this is a terrible question? Imagine that someone asked how we could measure your effectiveness as a parent. You might take offense at this question or you might say, "My kids love me and lead happy and fun lives." Or you might say how well they were doing in school. What you would probably not say is, "I am a very effective parent and I know this because my kids got 800s on their SATs."

But if these people were put in charge of parenting, or if they invested money in parenting classes or mentoring, they wouldn't accept subjective answers.

In their minds, effectiveness always means test scores. We don't have them yet for parenting, but we sure have them for school and that is because measuring effectiveness is now seen as a legitimate question to ask.

I always thought I was an effective teacher when my students got excited about ideas or came up with ideas of their own. But no. Now, we have to measure it, and in order to do that, we have to measure kids. Maybe these people think they are helping or maybe all investors think like this. I don't know. I just know it is questions like these that kill educational innovation.

2. What learning analytics are they gathering on their offering and/or what do they plan to collect and why? What metrics matter to them?

Do our graduates get jobs? That would be the question that mattered to me. But no. They would hate that answer. Do our students enjoy themselves? No. They would hate that answer as well. Do our students send their friends? Maybe that one. But, alas, they mean test scores. Of course the only real question is do our students find their passion?

Do they know what they want and know how to pursue it? But, no one asks those questions. How sad for kids.

3. How standardized are the assessments of the artifacts of their story-based approach vs. having an expert pronounce it "just like we do it in the real world"?

Can these people think about anything but measurement? Nope.

4. Have they considered a competency-based learning model?

Aha. The real question. *Could you guys just forget about helping people learn to do things and just concentrate on the important stuff which as we all know is fact memorization, formula application, and lots of math which is after all so easy to test.*

Want to know why our schools are a disaster? Ask investors what they want to measure next.

Now I am disgusted!

Usually I am simply angry about the state of American education. Now I am disgusted.

I attended a meeting of philanthropists last week whose main concern was math and science education in the U.S. People who had achieved "success" in math and science education were brought out to show their wares. They told of how they got math scores to be higher and how they helped kids "prepare for college." So, of course, knowing that all this rests on the assumption that drilling kids for tests is useful – when no adult actually could pass any of these tests – or that college life requires geometry or trigonometry, which it never does (or at least shouldn't), I was my usual angry self. But then…

The lunchtime event included 10 kids in uniform who were marched on stage and began to sing. A little entertainment I thought – too bad they look so uptight. But then the song…

They sang a song about multiplication. It seemed to go on forever and listed every multiple of every number up to 13 or so. The kids were very excited to be singing this song. Then the kids were asked questions.

Was this a good way to learn? A resounding *"Yes!"*

Did they like working longer hours and on Saturdays? *"Oh yes!"*

Did they like the system that this school employed that meant that no lesson was completed until the weakest student had mastered this lesson? *"Oh yes!"*

I wanted to ask them if they were allowed to express actual opinions in this school. I wanted to ask if whenever they actually multiplied did they have to sing the whole song to themselves? I wanted to ask them if they felt like killing the dumb kids and how they coped with the boredom of this school. I wanted to ask them if they realized that memorization, apart from the multiplication table, basically is unnecessary throughout life?

But, I didn't. I was too sick to ask anything.

This is an example of a very successful school. Why is it successful? Because these kids get into college at a better rate than their peers in other schools. The fact that this same effect would probably occur if these same kids were paid attention to and allowed to think for themselves was not mentioned.

We are producing robots and proud of it. More money for math and science test passing!

Chapter 7: Dear Mr. Obama (and Other Politicians)

Mr. Obama wants big ideas? Here are 10 in education.

At a fundraiser yesterday in San Francisco, President Obama said that "We have lost our ambition, our imagination, and our willingness to do the things that built the Golden Gate Bridge..."

No, Mr. President, it isn't "we" it is you. There are plenty of good ambitious ideas out there, you just aren't listening.

Here, off the top of my head, are ten outrageous big ideas about education. You will listen to none of them. You have considered none of them. You haven't even tried to understand them. Yes, they sound crazy, as do all new ideas.

Ten Big Ideas In Education

1. Shut down high schools
2. Stop preparing students for college
3. Stop insisting everyone go to college
4. Re-focus colleges away from academics
5. Eliminate all testing
6. Get big business out of education
7. Make learning fun again
8. Let children choose what they want to learn about
9. Help children find mentors who will help them learn what they want to learn
10. Build online experiences that engage students and that teach thinking skills

I have written about these ideas in more detail elsewhere and won't repeat myself here. Suffice it to say that a high school system designed for the elite in 1892 could not possibly be right-headed today. Yet instead of changing it you are making sure that we test every students to tears to make sure they have memorized the quadratic formula, disregarding the fact that hardly any adult actually uses it.

Re-think what you are doing in education, Mr. Obama. You have become the problem.

There are plenty of ideas out there.

Duncan speaks; kids lose

While the testing companies make great profits, the nation's newspapers, having a vested interest in those profits, tout testing as the country's salvation. The most visible touter is, of course U.S. Secretary of Education Duncan, who gives eloquent speeches that are, of course, printed in the Washington Post (who owns Kaplan Testing). Here is an excerpt from one of them given in September 2009, together with my comments:

> Let's build a law that respects the honored, noble status of educators – who should be valued as skilled professionals rather than mere practitioners and compensated accordingly.

Duncan is saying that teachers are wonderful people so therefore it follows that No Child Left Behind is a great law.

> Let us end the culture of blame, self-interest and disrespect that has demeaned the field of education. Instead, let's encourage, recognize, and reward excellence in teaching and be honest with each other about its absence.

Then he says that we should like teachers a lot because they will help raise test scores.

> Let us build a law that demands real accountability tied to growth and gain in the classroom – rather than utopian goals – a law that encourages educators to work with children at every level – and not just the ones near the middle who can be lifted over the bar of proficiency with minimal effort. That's not education. That's game-playing tied to bad tests with the wrong goals.

Then he says there should be accountability which is the code word for testing that makes it sound like it doesn't mean that number 2 pencils and bubble sheets are what education will be all about.

> Let us build a law that discourages a narrowing of curriculum and promotes a well-rounded education that draws children into sciences and history, languages and the arts in order to build a society distinguished by both intellectual and economic prowess.

Then he says that the curriculum should be exactly what it always has been and no other ideas will be accepted.

> Let us build a law that brings equity and opportunity to those who are economically disadvantaged, or challenged by disabilities or background – a law that finally responds to King's inspiring call for equality and justice from the Birmingham jail and the steps of the Lincoln Memorial.

Next he argues that black people should have good schools. Does anyone disagree with that? What is his plan? More testing.

> Let us build an education law that is worthy of a great nation – a law that our children and their children will point to as a decisive moment in America's history – a law that inspires a new generation of young people to go into teaching – and inspires all America to shoulder responsibility for building a new foundation of growth and possibility. I ask all of us here today – and in school buildings and communities across America -- to roll up our sleeves and work together and get beyond differences of party, politics and philosophy.

Next he argues that a good education law would encourage good people to become teachers. While that is true, he certainly isn't proposing such a law.

> Let us finally and fully devote ourselves to meeting the promises embedded in our founding documents – of equality, opportunity, liberty – and above all -- the pursuit of happiness. More than any other issue, education is the civil rights issue of our generation and it can't wait -- because tomorrow won't wait – the world won't wait – and our children won't wait.

Then he equates education with civil rights, which means mostly that he is looking to woo the black vote.

Impressively said. Duncan can sure talk. But the speech means nothing and flies in the face of reality. This is all a justification for continuing the policies of the Bush administration in education. Why would Obama want to do the same thing as Bush did especially when he campaigned against No Child Left Behind? The answer is simple. There has been lots of money invested in testing by powerful players that Obama doesn't want to offend. Sadly, the kids are noone's main concern.

A message to Bachmann, Duncan, and every other politician who thinks he or she knows how to fix education

Michelle Bachmann has an education agenda. All politicians have an education agenda. They all are sure the schools are broken. This leads to two obvious questions:

1. Why do they all agree the schools are broken?
2. Why are their solutions always to the left of insane?

As for the insanity question, bear in mind that this is simply not a matter of politics. Bush's policies in education were insane. Obama's policies are insane. And, all the people running against Obama have insane educational policies. Why is this? How can this be?

The obvious question is what is insane about them. To answer that we need to address question #1.

Here are some reasons we hear about why schools are broken (and my responses):

There is a lack of discipline
You try making 30 kids sit still all day, especially in the modern era.

The teachers are often not very good
There certainly are mediocre teachers but there are also some very good ones, which is amazing because it becomes more difficult each day to put up with the rigid system we have created for them to teach in.

The average American doesn't know ______ (Fit your favorite in here: who George Washington was, the capital of Delaware, where Iraq is on a map, the quadratic equation…)
Knowing facts really doesn't matter in any way. Because schools teach facts and test facts we have become convinced that facts matter. Facts that do matter in your life tend to be learned while doing (like the names of streets are learned by those who walk or drive on them). Otherwise it is knowing *how*, not knowing *facts* that matters.

Everyone needs to go to college and high school isn't preparing them properly
Everyone does not need to go to college. College as it exists today bases its curriculum

on a research model that is driven by faculty recruitment. Universities teach students to be researchers not practitioners. Even masters programs which are supposedly designed to train practitioners, tend to be dominated by theories and arcane subjects that will never matter to a practitioner. We need to move to a more practical notion of education that leads to jobs. Liberal Arts colleges eschew this notion. We can't afford many more Literature majors.

Tests scores in basic skills are bad

Tests are moronic. Yes, moronic. If the tests tested performance they might have some credibility, but multiple choice tests test nothing. Every driver who has to take a multiple choice test to renew his license has to study the manual first no matter how good a driver he may be. Multiple choice tests test only one's ability to prepare for and tolerate multiple choice tests.

We need citizens with 21st century skills and school isn't doing this

I am not sure what 21st century skills are but I am pretty sure they include reasoning, communication, and human relations, which were good in any century and are really not part of K-12 curricula. What we need is a populace who can think clearly, which, judging from the extant political candidates, we clearly do not have.

We need more scientists and engineers

We have plenty of scientists and engineers. If anyone thought we really needed more they would create a high school engineering curriculum. But that would mean throwing something out and the 1892 curriculum has become sacred.

There needs to be more religion in schools

Really? There needs to be religion in schools? Whose religion exactly? And why? So we can ram more facts into kids' heads. Facts are only the medium of education because religious institutions were the designers of the schools in the first place.

Schools don't teach everyone to love America enough

School should teach students to criticize America not love it. With thoughtful criticism comes change.

Schools are dangerous places

This last one is right. Schools are very stressful places and they are places where bullying happens and where kids learn to feel bad about themselves unless they have a really good teacher who can make sure none of that happens.

My message to Michelle Bachmann and Arne Duncan and all the other fools who pontificate about education is simply this. If we had a good education system, maybe you all could reason better and would stop saying and doing insane things about education.

Just the facts, ma'am

Another brilliant revelation from our heroes in Washington:[1]

> Students who complete Algebra II are more than twice as likely to graduate from college compared to students with less mathematical preparation.

Would you like to know why this is true (and I have no doubt that it is true)? The answer is given further down in the article:

> The report also cited findings that students who depended on their native intelligence learned less than those who believed that success depended on how hard they worked.

See, this is an easy one. If you work harder you get into college. Now the question is: why are we making the thing that students have to work harder at – Algebra II?

We know why this panel decided that. At stake is a $100 million federal budget request for math. Now and guess who was on the panel?

I dunno. People who might receive that funding would you guess? You betcha. A panel of university folks who are just dying for that grant money to be approved worked on a very well-funded study that proved that the nation would not succeed without that grant money.

My favorite part of the *New York Times* article was the following:

> Dr. Faulkner, a former president of the University of Texas at Austin, said the panel "buys the notion from cognitive science that kids have to know the facts."

No, Dr Faulkner, as a graduate of your esteemed institution, and as a founder of the field of cognitive science, let me suggest, with all due respect, that it is you who needs to know the facts.

The first fact is that you are a chemist, and I am pretty sure you don't really know much about cognitive science.

The second fact, is that there is plenty of work in cognitive science that shows that background knowledge helps one interpret the world around one, and thus reading, for example, is facilitated by understanding something about the world you are reading about.

The third fact is that there is no evidence whatsoever that accumulation of facts and background knowledge are the same thing. In fact, there is plenty of evidence to the

1 Report Urges Changes in Teaching Math http://www.nytimes.com/2008/03/14/education/14math.html?pagewanted=print&_r=0

contrary. Facts learned out of context and apart from real-world experience that is repeated over and over are not retained.

The fourth fact is that kids don't like math much and it is clear why. They find it boring and irrelevant to anything they care about doing. If you think math is so important, then why not teach it within a meaningful context, like business, or running a school doing the kind of math you had to do to do that – which certainly wasn't Algebra II. There is plenty of evidence that shows that teaching math within a real and meaningful context works a whole lot better than shoving it down their throats and following that with a multiple choice test.

The fifth fact is that there is no evidence whosoever that says that a nation that is trailing in math test scores will somehow trail in GDP or whatever it is you really care about. This is just plain silly, but we keep repeating the mantra that we are behind Korea in math as if it has been proven that this matters in some way. Nothing of the sort has been proven.

The sixth fact is that there are lots of vested interests that need to keep teaching math. Let me name them: tutoring companies, testing companies, math teachers, book publishers, and many others who make lots of money when people are scared into thinking that their kid won't get into college because he or she is bad at Algebra II.

The seventh fact is that nearly every grown adult has forgotten whatever algebra he or she ever learned to pass those silly tests, so it is clear that algebra is meaningless for adult life. I ask every important person in public life that I meet to tell me the quadratic formula. No one has ever been able to do so.

The eighth fact is that any college professor who is honest will tell you that algebra almost never comes up in any college course, and when it does come up it usually needn't be there in the first place.

I know this is a hopeless fight, but algebra really matters not at all in real life and the country will not fall behind in any way if we simply stop teaching it. That is not a fact, it is just a former math major's, UT graduate's, and computer science professor's point of view.

Thank you Arne, Bill, and Pearson for making this teacher so miserable

Many people write to me, especially unhappy students and teachers who have had it with the system in which they teach. Here is a letter I received a few days ago:

> I believe as you do that stories have power. I teach history where story should play a more prominent role in our curriculum. I have been a teacher for 11 years. While never a strong supporter of the public school system, I am a strong supporter of public education. The system is killing the hope of everything this country could possibly achieve. Public education is truly in critical condition. I am especially concerned for the minorities and those in poverty. We do not have an equitable system, but we expect equal results.
>
> I read your book, Teaching Minds, with great interest. I love studying about curriculum and cognitive science. Education, in general, is my passion. Teaching seems to be a natural outgrowth of this, but it has not been as enjoyable as I had hoped it would be. I have a Bachelor's degree in Interdisciplinary Social Science with a minor in Psychology and concentration in History. I currently teach US History at the high school level. I also have a Special Ed endorsement. Most of my teaching has been in Special Education classrooms. I now have a Master's in Curriculum and Instruction. I constantly think about how to create curriculum that makes sense. Hence, your book was like taking a deep breath after being submerged in an ocean of chaos and confusion.
>
> Like you, I grew up hating school, but loving to learn, and when my children were born I embarked on my adventure of learning about education in the hope of keeping their love of learning alive. I dabbled in homeschooling them for a time. When they did go to school, I supplemented with a variety of experiences and believed in "unschooling." Eventually, I began to get my degree in education with the dream of establishing an alternative school. Now, over a decade later, I still dream of such a school, but have found myself stuck in the mire of our public school system.
>
> I want to engage students, motivate them to learn and be self-disciplined (another skill that desperately needs to be learned). I would love the feeling that I had actually done a good job when I fully retire. Following administration's guidance has only led me to feel less competent and less effective than ever. We all know we should lecture less, if at all, but what we have to replace it with is worksheets and graphic organizers that mimic the ACT. I am required to give practice ACT tests throughout the year. To counteract that, we also have Document Based

> Questions that are supposed to encourage critical thinking, but the kids still don't care. I have lost all student engagement. It has been most disheartening.
>
> I would love to have the opportunity be part of Alternative Learning Places. It is exactly what I have been dreaming of. I plan to try and implement your ideas into my curriculum next year, if administration allows me (or I manage to sneak it in). Next year is not an evaluative year for me and I am close to retirement age, so I may go rogue. At this point, I have nothing to lose except the boredom of my students. I think gaining teacher support would be easier than gaining administrative support and, if we banded together, I believe we could make things happen. I am willing to help on that score, as well. Teachers want to make a difference in their students' lives. They want their students to want to learn. They don't want to work hard with nothing to show for their efforts and then be blamed for the outcome of something over which they have no control.
>
> Thank you for your book and your courage to share your ideas. I stand with you in the hope changes can be made. It's time for a revolution!

I found this letter extremely difficult to read. How miserable we have made teachers (and students). Why does someone who really seems to care about her kids being excited about learning have no real way to do that?

Thank you Arne Duncan. Thank you Bill Gates. Thank you Pearson Publishing, McGraw Hill, ACT, ETS, and all the other organizations who just want a world where there are tests to take and teachers to make sure students take them. Thank you for making it nearly impossible to make any changes because of Common Core and because of your tests. Thank you for making teachers miserable by judging them by how their students do on your tests.

I don't know what these people's real goals are (except making more money). I have trouble believing they just hate kids and hate teachers. But they sure don't care about letting kids have fun learning and letting teachers have fun teaching people who are excited to learn.

Why are you proud, Mr. Mayor?

Today the *New York Times* published this editorial, "Good News on Math."[2]

> Mayor Michael Bloomberg and the teachers of New York are rightly proud of the city's performance on this year's state math tests. New York City students showed gains in every grade tested, outpaced students in most other of the state's big cities and edged closer to the state performance average. The new scores, which showed that 65.1 percent of city students are performing at or above grade level — are up from 57 percent last year. The news is not as good for the city's eighth graders. Only 45.6 percent of them were found to be proficient in math. These disappointing results suggest a need for stronger instruction in the sixth grade, where students may not be getting the skills they need to master more complex, middle school material.

And now for my questions:

1. Why is Mayor Bloomberg proud?
2. How would Mayor Bloomberg do on these math tests?
3. How would the writer of this editorial do on these math tests?
4. Why is the goal to prepare students for more complex middle school material?
5. What does mastering middle school math prepare students for?
6. Assuming the answer to #5 is high school material, what does that prepare students for?
7. Assuming the answer to that is college material what does that prepare kids for?
8. Given that most students never use the mathematics they learn in high school ever again the rest of their lives, why are we playing this silly game?
9. Could the answer to that be that Mayor Bloomberg wants to be able to say he did something important in education even if by any reasonable standard he clearly didn't?
10. Since when does 35% of students failing constitute success at anything?
11. If every student in New York were good at mathematics in what way would our society be better off?
12. Why is the goal to beat other cities and states?
13. Is New York in competition with other cities and tests in some math contest we don't know about?
14. What good happens in New York if it wins that competition?
15. What good happens to Mayor Bloomberg if New York wins that competition?

2 http://www.nytimes.com/2007/06/15/opinion/15fri3.html

16. Why does the *New York Times* care about any of this?
17. Does the *New York Times* realize that every time they crow about nonsense such as this they make mathematics more and more important in the curriculum?
18. Are all the people at the *New York Times* experts in mathematics?
19. If they are experts in something else, like say writing, thinking, working at deadlines, preparing coherent reports, reasoning about hard political problems, and such, why wouldn't those be important parts of the curriculum?
20. Could it be that when we emphasize mathematics we de-emphasize the very things the people at the *New York Times* are good at?
21. Does anyone care that the system is now totally insane?
22. Does the *New York Times* realize it is making matters worse in education with editorials like this?
23. Does Mayor Bloomberg or the *New York Times* actually care about education?

More college graduates? Say it ain't so Mr. President.

Mr. Obama has promised that by 2020, America will "once again have the highest proportion of college graduates in the world."

Can we think about that for a minute? Why does this matter? We Americans so believe in college that we rarely ask why. I asked my students at Yale and later at Northwestern why they were in college. I heard a lot about parties, a four-year vacation, going because their parents made them come, and a lot about how you need a college degree to get ahead.

I never once had anyone say they were there to learn. Never.

College offers diplomas. Education, not so much. And to the extent that colleges do offer education (a Yale English major is typically considered well-educated by modern standards), what difference does it make to the country? Those English majors don't easily find careers and in bad economic times even a Yale degree may not buy you much.

I am sure what Mr. Obama meant to say was that by 2020 our population will be able to reason effectively, work well with others, and communicate well. At least that is something he quoted from me while he was campaigning. But alas, now it all about making sure those 3,000 colleges we have survive regardless of whether they are turning out more productive and reasoning citizens.

Yes, Mr. Obama, money is the answer

Yes, Mr Obama, money alone can solve the school predicament. Use the money to create new curricula that are both relevant and interesting. Stop teaching the "scholarly subjects" determined by the Harvard President in 1892. We don't need a nation of scholars. The kids know that, so they ignore what teachers are forced to teach. Try teaching students to get good at doing what they find interesting to do. Indeed, money could allow that to happen. Also, perhaps you could try fighting the special interests that want to preserve the testing culture.

What were your test scores, Mr. Obama?

"It means treating teachers like the professionals they are while also holding them more accountable," Mr. Obama said. "Good teachers will be rewarded with more money for improved student achievement."

Really? Is that how professionals are treated these days? Do we measure other professionals by how those they mentor do on standardized tests? Would you, Mr. Obama, like to be measured by how your staff does on standardized tests?

Treating teachers like professionals might include letting them actually teach to a student's interest and concerns rather than helping them raise their math scores.

And, while we are at it Mr. Obama, what were your scores on the SAT? Did anyone ask you that while you were campaigning? Would you have thought it was stupid if they had?

Tell me the quadratic formula Mr. Obama. Can't? Might that be because it doesn't matter in any way to know it? Stop making kids memorize nonsense and you will be treating teachers like professionals.

The school of random facts

I happened on an article in *The Huffington Post* written by someone named Schweitzer who is listed as "having served at the White House during the Clinton Administration as Assistant Director for International Affairs in the Office of Science and Technology Policy." Here is a piece of what he said:[3]

> The health care debate cannot be understood in historic context because many Americans have never heard of Thomas Jefferson. Extrapolating from state surveys, only 14% of American high school students can name who wrote the Declaration of Independence. Nearly 75% do not know that George Washington was our first president... We can say that our educational system has failed when the vast majority of American students do not know enough to pass an exam to qualify as American citizens.

Really?

First let's talk about why we have such a failed system. Could it be the policies of Presidents like Clinton, who pursued a policy of never offending the teacher's unions by doing anything threatening to them like changing the curriculum?

Or, could it be that fools like you define education in terms of random facts you wish everyone knew? The problem is not that people don't know who Thomas Jefferson was. If citizens knew who he was would that mean that they can now think clearly and not be influenced by all the special interests who are trying to tell them what to think? If they knew who George Washington was, what exactly would they know about him? *That he could never tell a lie?* Obviously untrue. *That he was a brilliant general?* Doubtful. *That he owned 300 slaves?* Not usually mentioned. *That he married a rich woman so he could get her land?* Nah. You are upset because our students don't know our national myths and some random facts.

I am upset that people can't think clearly. Surely this is the problem with our so-called national debate on health care. Surely the schools could address this issue. Nah, it would mean giving up on tests that see if students can memorize a right answer.

3 A Failure of Citizenship and the Health Care Debacle http://www.huffingtonpost.com/jeff-schweitzer/a-failure-of-citizenship_b_291539.html

Hooray for the Democrats! Hooray for more accountability!

The following is from an article on the front page of the *New York Times*, "Democrats Make Bush School Act an Election Issue":[4]

> Mr. Obama, for instance, in a speech last month in New Hampshire denounced the law (NCLB) as "demoralizing our teachers." But he also said it was right to hold all children to high standards. "The goals of this law were the right ones," he said.
>
> When Mr. Edwards released an education plan earlier this year, he said the No Child law needed a "total overhaul." But he said he would continue the law's emphasis on accountability.
>
> And at the elementary school in Waterloo, Mrs. Clinton said she would "do everything I can as senator, but if we don't get it done, then as president, to end the unfunded mandate known as No Child Left Behind."
>
> But she, too, added: "We do need accountability."

Accountability must play well in Peoria because every Democratic candidate is simultaneously for it while being against NCLB.

The question is: how can you hold both positions?

Here is how. *By not understanding the issue.*

Accountability must mean to voters, I assume, that teachers will be measured by how well they teach their students. Those fearless Democrats, always willing to hop on an uncontroversial point of view, are all quite certain that the voters know what they are talking about. No matter how stupid NCLB is, no matter how mean spirited, no matter how awful for both teachers and students, its very horror rests on the premise that no one seems to be disputing, that the federal government has the right to tell the schools what to teach and to see if they are indeed teaching it.

How is this premise wrong? It's based on false assumptions. Let's take them one by one.

All schools should teach the same subjects

Why is this wrong? Because kids in New York come from, and will live in, a different world than their compatriots in New Mexico. In New Mexico, I was asked if we could teach Casino Management and Land Use. Yes, we could, but not if there is federal accountability about algebra and twenty other subjects that make it impossible to fit these subjects in.

---

4 http://www.nytimes.com/2007/12/23/us/politics/23child.html?pagewanted=print

There is no right set of subjects. The fact that the President of Harvard in 1892 thought there were and thought he could say exactly what they would be in the 21st century does not make it true. (OK, probably he wasn't thinking about the 21st century in 1892, but we all seem to think he must have been because we are still teaching the same stuff.)

Some subjects are more important than other subjects

Yes, we have electives. But they don't matter. Because accountability means making sure that we teach what does matter first. What matters? The stuff that we are holding people accountable for. Since this seems to be math and science these days, for no good reason I can discern, this means that we will get to the stuff that would excite kids and keep them in school and, horrors, might teach them some job skills, after we are done with the important stuff. Sorry candidates. I absolutely guarantee that none of you know the quadratic formula or the elements of the periodic table which is of course, the stuff of accountability since it is so easy to test. Then, how can that be the important stuff? How about how to see what voters are thinking and then say it to get elected? That is the important stuff in your lives. Why not teach that?

All important subjects can be easily tested

Yes, there are right answers in math. But are there right answers in whether we should invade Iraq? No? Does that mean we can't teach how governments actually work and how to get reasoned arguments to be heard? Is there a right speech candidates should make? Does that mean we can't ask students to give speeches because we can't easily assess them? Do we only teach subjects for which there are clear right answers? We do now, which is one reason why school is a deadly experience for one and all and will remain so as long as accountability is the key word in government.

Seeing who did better than whom in school is an intrinsic part of the educational process. Admit it candidates. It really is all about competition isn't it? You are all the winners of the school competition. You went to Ivy League schools and did well. Well, hooray for you. I taught at Ivy League schools and I was profoundly unimpressed with the test-taking, grade-grubbing students I found there. The goal of education is not to say who won and it is not to tell Harvard whom to admit. The goal is provide real-world skills, some of which may not be so easy to assess until the graduate actually shows up in the real world.

All children have the same educational needs

There is a 50% dropout rate in many high schools because we have forgotten that not everyone is going to Harvard and that going to Harvard is not the goal of education. Some children simply need to learn about ethics and business and child-raising and how the legal system works, how to take care of their health and how to understand when politicians are saying things that make no sense.

Why wouldn't those subjects be critical? I bet not one of you thinks any of those are more important than math and science. How about the student who has a passion for the environment, or doing social good, or being a good parent, or, perish the thought, running for office? Couldn't we teach those subjects simply because students have said they want to learn them? Does every school have to be the same?

Needless to say, I have some problems with these assumptions and so should the Democratic presidential candidates. I can excuse the voters for not understanding these issues, but I will not excuse President Bush and his cohorts, who I sincerely doubt give a hoot about education, nor will I excuse the Democratic challengers who should know better.

I have an idea. Why not just keep the federal government out of the education business and simply leave schools alone? Educators have enough trouble fighting the silly standards that colleges impose upon them without having to put up with whatever version of accountability you choose to proffer after your election.

Why politicians and rich guys won't reform education

I am glad that some rich guys have decided it is important to fix American education. Their method – get the candidates to take education reform seriously – sounds good on the surface but it will be another waste of money – the same money that could actually fix the nation's schools.

How do I know this? I know why politicians will never seriously tackle education reform. Here are the constituencies they would have to antagonize in order to get reform to actually happen:

Teachers – Teachers would have to teach differently and no one really wants to change what they do on a day-to-day basis. True, teacher's lives have been made so miserable by previous politician's attempts at reform that they are more open to change than ever, but still, they really don't want to have to go to school to learn new methodologies.

Publishers – Big corporations have a real stake in education staying the way it has been. They don't want to throw out all their textbooks and start over. They spend a lot of money making sure this doesn't happen.

Testing companies – Politicians have helped create an enormous industry that prepares and grades tests. They won't give up their business without a fight. No real reform will take place if teachers are still teaching to the test and if we continue to teach stuff that is easy to test rather than giving kids open-ended issues to think about and real workplace skills.

Universities – Any real school reform means changing how universities conduct admissions and convincing them to teach subjects in college they have foisted upon the high schools (like algebra). This will never happen since it would also mean that colleges would need to interview students instead of relying upon grades and test scores for admission.

Parents – Parents tend to think school is a competition and they reinforce all the testing and grading in the hopes that their kid will win. In addition they believe that whatever they learned in school is what should be taught despite the fact that they have since forgotten all that they learned in school.

The real question is deciding what school is about. John Adams said that school is about learning to make a living and learning to live. George Bush thinks school is about passing tests (not that he could pass them).

To rich guys – Help build an alternative to the present school system with your money. The current system simply cannot change. There are too many vested interests in the status quo. No one who is running for President will ever demand substantive change.

Public schools: where poor kids go to take tests

This news appeared today in the Washington Post:[5]

> For the first time in at least 50 years, a majority of U.S. public school students come from low-income families, according to a new analysis of 2013 federal data, a statistic that has profound implications for the nation.
>
> The Southern Education Foundation reports that 51 percent of students in pre-kindergarten through 12th grade in the 2012-2013 school year were eligible for the federal program that provides free and reduced-price lunches.[6] The lunch program is a rough proxy for poverty, but the explosion in the number of needy children in the nation's public classrooms is a recent phenomenon that has been gaining attention among educators, public officials and researchers.

Why is this true do you think? Seems simple enough. If you can possibly afford it you wouldn't even think about sending your kid to public school unless there happened to be a safe school with interesting and fun teachers who did exciting things in a public school nearby. And what are the odds of that?

Thank you Mr. Bush, and Mr. Obama, and especially Mr. Duncan, for making school even worse than it was before by having a policy of constant testing to see how everyone is doing. Under the guise of helping poor people do better you have pushed richer people out of the system. No one wants to use your public schools. Try thinking about that the next time you make more standards that make school a nightmare of test preparation and testing.

The latest salvo was from Mr. Obama and his henchman Tom Hanks, trying to convince everyone that it is OK for high school to be an awful experience because you can go to community college for free and that will solve everything, The *New York Times* printed that and I am guessing that Obama's staff wrote it. They will do anything to avoid the obvious conclusion that the schools aren't working.

Here is a simple idea: let people who want to make changes in high schools make them. We can teach job skills, life skills, and make it fun. Or, we could make all the poor people learn algebra, chemistry, and history so they can remain poor having learned nothing of use to them.

---

5 Majority of U.S. public school students are in poverty http://www.washingtonpost.com/local/education/majority-of-us-public-school-students-are-in-poverty/2015/01/15/df7171d0-9ce9-11e4-a7ee-526210d665b4_story.html

6 A New Majority Research Bulletin: Low Income Students Now a Majority in the Nation's Public Schools http://www.southerneducation.org/Our-Strategies/Research-and-Publications/New-Majority-Diverse-Majority-Report-Series/A-New-Majority-2015-Update-Low-Income-Students-Now

Duncan and Obama are actively preventing meaningful education change

Bill Maher made an important statement last week when he criticized President Obama for failing to do much to satisfy those who had voted for him.

> But when I read about how you sat on the sidelines while bailed-out banks used the money we gave them to hire lobbyists who got Congress to stop homeowners from getting renegotiated loans, or how Congress is already giving up on health-care reform, or how scientists say it's essential to reduce CO2 by 40% in 10 years, but your own bill calls for 4%, I say, enough with the character development, let's get on with the plot.

But Maher left out education. How is our President doing on education? His education secretary has announced plans for national standards:

> This Sunday Duncan proposed a system in which schools signing on to the standardized benchmark will benefit from a $350 million pot aimed at assisting in the development of the new test needed to measure the potentially nationalized educational standards.

He has also announced his intention to lengthen the school day and school year with the express intent of helping students prepare for tests.

Or, let's put this another way. The President has so failed on education already that his failures on banking, health care, and the climate pale by comparison, He has effectively said to people like me, who are actually trying to make real change in the schools: forget your ideas about teaching modern subjects (like scientific reasoning, medical decision-making, internet startups, or how to take care of a child) because we are going to continue to ram the algebra, U.S. history, and science facts view of education down everyone's throats and will actively prevent meaningful change. Forget learning by doing – there will be none of that. There will learning by memorizing.

WHY?

I asked my sources in the White House. "We are trying to get re-elected here. The voters care about test scores." That is what they said. Really.

Bill Maher, you don't realize how bad it really is.

Free community college? How about we fix high school, Mr. Obama?

The *New York Times* explained this morning what is behind the "free community college plan" of President Obama. In this article they said:[7]

> The United States built the world's most successful economy by building its most successful education system. At the heart of that system was the universal high school movement of the early 20th century, which turned the United States into the world's most educated country. These educated high school graduates — white-collar and blue-collar alike — powered the prosperity of the 20th century.

That may well be true. The high schools of the early 20th century taught employable skills (in addition to the absurd 1892 academic curriculum still in place). Eventually all practical high school programs were eliminated from high school because everyone "must go to college."

Mr. Obama, instead of restoring all the practical things that were taught in high school, wants to make everyone go to college in order to learn employable skills.

> The plan would allow anyone admitted to a community college to attend without paying tuition, so long as they enroll in a program meeting certain basic requirements and they remain on track to graduate in three years. Its broad goals are clear: to extend the amount of mass education available, for free, beyond high school — from K-through-12, to K-through-college. "The president thinks this is a moment like when we decided to make high school universal," said Cecilia Muñoz, director of the White House Domestic Policy Council.

Here is a wild suggestion, Mr Obama. Fix high school. Teach practical subjects there. Eliminate the 1892 curriculum.

Here are some suggestions for what could be taught in high school today:

7 The Roots of Obama's Ambitious College Plan http://www.nytimes.com/2015/01/09/upshot/the-roots-of-obamas-ambitious-college-plan.html

Some Proposed High School Curricula

Criminal Justice	Sports Management	The Music Business
Music Technology	Law	The Legal Office
Military Readiness	The Fashion Industry	Electrical Engineering
Civil Engineering	Robotics	Computer Engineering
Computer Networking	Homeland Security	Medicine
Nursing	Medical Technology	Construction
Television Production	Real Estate Management	Landscape Architecture
Computer Programming	The Banking Industry	The Investment World
Automobile Design	Aircraft Design	Architecture
Biotechnology Lab	Film Making	Travel Planning
Financial Management	Accounting	Parenting and Childcare
Animal Care	Zoo Keeper	Urban Transit
Hotel Management	Healthcare Industry	Food Industry
Graphic Arts		

Could we do this? Easily. Online education allows teaching anything anywhere. Every kid could choose what they were interested in and then change his or her mind and do something else if they got interested in something else. And there are many more possibilities. Spend our money more wisely Mr. Obama. Build that.

Community college wouldn't be necessary if the high schools weren't broken.

Chapter 8: The New York Times (and Others) - Wrong Again

The New York Times and Nick Kristof want mass education. I want individualized education.

Sometimes when I read the *New York Times* on education, I find myself wondering if they just sit around and think how they can write dumb stuff.

On Sunday, Kristof wrote a column that included this:[1]

> Until the 1970s, we were pre-eminent in mass education, and Claudia Goldin and Lawrence Katz of Harvard University argue powerfully that this was the secret to America's economic rise. Then we blew it, and the latest O.E.C.D. report underscores how the rest of the world is eclipsing us.
>
> In effect, the United States has become 19th-century Britain: We provide superb education for elites, but we falter at mass education.

We were pre-eminent in mass education and now we are not. Here is why we were pre-eminent in mass education. Our system was designed to train the masses to be factory workers. In 1905, Elwood Cubberly, the future Dean of Education at Stanford, wrote that schools should be factories:

> ...in which raw products, children, are to be shaped and formed into finished products...manufactured like nails, and the specifications for manufacturing will come from government and industry.

William Torrey Harris, US Commissioner of Education from 1889 to 1906, wrote:

> The great purpose of school can be realized better in dark, airless, ugly places.... It is to master the physical self, to transcend the beauty of nature. School should develop the power to withdraw from the external world.

And now, we don't have any factories. Should we still strive to lead the world in mass education?

Kristof also wrote this:

> The United States is devoting billions of dollars to compete with Russia militarily, but maybe we should try to compete educationally. Russia now has the largest percentage of adults with a university education of any industrialized country — a position once held by the United States, although we're plunging in that roster.

1 The American Dream Is Leaving America http://www.nytimes.com/2014/10/26/opinion/sunday/nicholas-kristof-the-american-dream-is-leaving-america.html?_r=0

We have been pushing everyone to go to college in this country for many years. The result is that college has become about football, partying, and suffering through lectures to accumulate credits for the degree. It also leaves students with large amounts of debt. It is mass education all right. 1,000 people crammed into a lecture hall is mass education, except it isn't education at all really.

What the US should strive to do is lead the world in – **individualized education**. I am sure Russia is good at treating everyone as a cog in the wheel of the great machine. Maybe that is even good for their economy. I don't know. But mass education is a terrible thing to be hoping for. We have no more factories and what students learn in college usually does not render them particularly employable.

On either hand we have the ability to do individualized education now. We can match mentors to students online. We can offer courses chosen by students as opposed to ones required by faculty. We can help students learn what they want to learn when they want to learn it. We can help employers find employees by allowing them to offer education that leads to employment. The one-size-fits-all concept of education that has dominated the US for the last 125 years needs to go.

To create individualized education, we need to start spending money on it. We can and should build all kinds of learn-by-doing experiences for children so they can try them out and see what they might like to do. We must stop shoving the old curriculum down everyone's throats and stop assuming that education for the masses is actually a good thing.

More from Kristof:

> In effect, the United States has become 19th-century Britain: We provide superb education for elites, but we falter at mass education.

What the elites that Kristof refers to in his article have is the opportunity to get some individualized education. That opportunity should be available to all, but it won't be as long as we spend money on mass testing instead of new curricula and new ways of teaching.

The *New York Times* just seems to love the word *massive*. They have been touting MOOCs which are just lectures without a professor around to talk with. Apparently the *Times* now wants to make sure that the masses are sufficiently educated. The usual reason for mass education throughout history has been to prevent revolution. The Communists and Nazis were very good at mass education.

Individualized education, Nick. It's coming. Ask any homeschooler.

Spinning test prep into "choice"

The *New York Times*, in yet another of its front page articles extolling improvements in education is very excited that: "Starting this fall, the school district in Chappaqua, N.Y., is setting aside 40 minutes every other day for all sixth, seventh and eighth graders to read books of their own choosing."[2]

Woo hoo!

You mean occasionally they will allow children to do something that they are actually interested in doing in school?

Not so fast.

Will students be able to bring in *Popular Mechanics* or even the *New York Times*? No, of course not. They will choose books approved by teachers. But, even this appalls the *Times* approved Bush appointee Diane Ravitch, who is always on the side of everything backward in education. She worries that no "child is going to pick up Moby Dick."

Indeed.

The *Times* goes on to say that:

> In the method familiar to generations of students, an entire class reads a novel — often a classic — together to draw out the themes and study literary craft. That tradition, proponents say, builds a shared literary culture among students, exposes all readers to works of quality and complexity and is the best way to prepare students for standardized tests.

It didn't take them long did it? Yea tests.

This is just more baloney intended to make the public feel like things are getting better in schools when In fact things are so bad that no one is happy (except maybe Diane Ravitch). You can't allow real choice in school because then you can't test it to see what kids have done.

I once built a program meant to get kids to learn the geography of the U.S without really trying, as they searched around the country for stuff they were interested in. It worked quite well. Kids loved it and they learned geography.

Nope. Rejected. Why?

---

2 A New Assignment: Pick Books You Like http://www.nytimes.com/2009/08/30/books/30reading.html?pagewanted=all

Because some students might go to California and others might go to New York. How would we test them? As soon as the tests appear innovation goes out the window. You mean kids would learn different stuff? Omigod!

In any case, this "choose what to read program" is an illusion. It is better than being force-fed Moby Dick for sure but what it is the real goal? The *Times* says; "Letting students choose their own books, they say, can help to build a lifelong love of reading."

That is the goal. Making kids read a lot in the hope that some of them will like it. Same as the math goal of shoving algebra down their throats in case anyone likes it. Kids rarely like what you make them do, or am I the only who has noticed that?

Can you live a long and happy life without having a love of literature? I think so. It is important to learn to read but that does not mean, by any means, that one needs to read "literature." If it isn't obvious to people by now, literature will soon be ancient history anyway. While humans have always told stories and always learned from them, they have not always had "literature." Novels have become commonplace for a very brief moment in human history and are now clearly being replaced by television and movies (for better or for worse, that is what is happening).

Teachers and politicians hate this of course. What I hate is that the idea of discussing life choices and issues in getting along in this world, which is a positive benefit of discussing literature, can only be done by reading Moby Dick according to the experts. There are many other ways to learn to think about life.

As a society, we have lost the forest for the trees. While we could be teaching deeply about why they do what they do, instead we are teaching them to pass tests. We insist that they learn what was fashionable for the elite to learn a century ago. We torture them and wonder why they drop out.

Moby Dick indeed!

The New York Times on the GED - wrong again

The *New York Times* is at it again, promoting nonsense about education. I can only guess that it owns a test-making or grading company because it sure does love this stuff.[3]

> Millions of Americans are trapped at the margins of the economy because they lack the basic skills that come with a high-school education. This year, more than 600,000 of these people will try to improve their prospects by studying for the rigorous, seven-hour examination known as the General Educational Development test, or GED, which should end in a credential that employers and colleges recognize as the equivalent of a diploma.
>
> The most fortunate live in states — such as Delaware, Kansas and Iowa — that have well-managed programs in which 90 percent or more of the test-takers pass.
>
> The least fortunate live in New York State, which has the lowest pass rate in the nation, just behind Mississippi. Worse off still are the GED-seekers of New York City, which has a shameful pass rate — lower than that of the educationally challenged District of Columbia. This bodes ill for the city, where at least one in five adult workers lacks a diploma, and the low-skill jobs that once allowed them to support their families are dwindling.

The *New York Times* wants to make sure that New York City has great GED courses so poor people can get jobs. For fun, I looked at the web site of the GED testing service. Here are three typical sample questions on a GED:

> "We hold these Truths to be self-evident, that all Men are created equal, that they are endowed by their Creator with certain unalienable Rights, that among these rights are Life, Liberty, and the pursuit of Happiness."
>
> Which of the following political actions violated the principle of "unalienable Rights" of liberty that evolved from the above excerpt of the U.S. Declaration of Independence?
>
> 1. In 1857, a U.S. Supreme Court ruling promoted the expansion of slavery in U.S. territories.
> 2. In 1870, the Fifteenth Amendment to the Constitution outlawed the practice of denying the right to vote because of race, color, or previous condition of servitude
> 3. In 1920, the Nineteenth Amendment to the Constitution granted women the right to vote nationwide.

3 How to Flunk Test-Giving http://www.nytimes.com/2009/10/13/opinion/13tue1.html

> 4. In 1964, the Civil Rights Act outlawed racial discrimination in employment and public accommodations.
> 5. In 1971, the Twenty-sixth Amendment to the Constitution extended the right to vote to 18-year-old citizens.
>
> A cook decides to recover some table salt that has been completely dissolved in water. Which of the following processes would be the most effective method of extracting salt from the solution?
>
> 1. spinning the solution in a mixer
> 2. boiling away the water
> 3. pouring the solution through cloth
> 4. dripping the solution through a paper filter
> 5. bubbling oxygen through the solution
>
> In May, I graduated from Prince William Community *College. Graduating with* an associate of arts degree in horticulture. Which is the best way to write the italicized portion of these sentences? If the original is the best way, choose option (1).
>
> 1. College. Graduating with
> 2. College, I graduated with
> 3. College. A graduation with
> 4. College. Having graduated with
> 5. College with

I don't know about you, but as an employer I know that I would certainly hire people for low-paying jobs if only they could answer these important questions. Perhaps it is time for the *Times* to notice that employers won't hire people who can't do anything useful and that our education system doesn't teach much that is useful. Pouring money into test passing courses will fix nothing.

Here is the *Times* again:

> New York will need to invest a great deal more than it spends at the moment. But the costs of doing nothing clearly outweigh those of remaking a chaotic and ineffectual system.

Right you are *Times*. New York needs to invest in real education however, not in test prep courses. How is it that the *Times* is this much out of touch?

When the New York Times obsesses about math, every kid loses

I have a confession to make. I did graduate admissions in computer science for more than 25 years. The first thing I looked for was the applicant's math GRE score. I eliminated anyone under 96th percentile. (Also, to add to my confession, I majored in mathematics in college.)

Why did I use this measure? Because ability to reason mathematically is an indicator of rigorous thinking, exactly the same kind of reasoning needed in computer science. Does that mean I needed my students to know mathematics? Not at all. Mathematics never came up in any way in our PhD program.

I mention this because there seems to be a national obsession with teaching mathematics and with math test scores. This is especially true when one reads the *New York Times*. Here are two articles published just this week:

"Don't Teach Math, Coach It"[4]

"Why Do Americans Stink at Math?"[5]

The first article is by a math professor who wishes his kid liked math as well as he likes baseball. It has no business being in the *Times* except that the *Times* seems a bit obsessed with math. The second article is really about how to teach better but the math panic headline is obviously exciting stuff to the *Times*.

Here is one more: "Math Under Common Core Has Even Parents Stumbling"[6]

Naturally the *Times* draws the wrong conclusion from their own article. Instead of realizing that Common Core math is out of the scope of even the parents of their own children, it goes on about teaching methods so that kids will do better at Common Core.

And the of course, we have the real stuff: "American 15-Year-Olds Lag, Mainly in Math, on International Standardized Tests"[7]

4 http://www.nytimes.com/2014/07/25/opinion/dont-teach-math-coach-it.html?gwh=36CC79FBFEE7D0518AF9F3D6D5B0B6F5&gwt=pay&assetType=opinion

5 http://www.nytimes.com/2014/07/27/magazine/why-do-americans-stink-at-math.html

6 http://www.nytimes.com/2014/06/30/us/math-under-common-core-has-even-parents-stumbling.html

7 http://www.nytimes.com/2013/12/03/education/american-15-year-olds-lag-mainly-in-math-on-international-standardized-tests.html

There is an international math competition going on and the *Times* wants the U.S. to win. We also want to win the bobsled competition.

It is time to be honest about what is really going on. Why is math important? *It isn't.*

(Now all the math teachers can tell me I am crazy, as they usually do.)

Why are math test scores important? All you need to know is here:[8]

> Harvard College announced Thursday that it has accepted 2,023, or 5.9 percent, of 34,295 students applying for admission to the Class of 2018.

I didn't really want to read 300 applications for computer science when I did admissions. So I took the easy way out. I relied on a simple but reliable metric. If you do well on an absurd math test, then it means you will work hard and can think logically, both of which are important qualities for a computer science PhD student.

Harvard cannot read 35,000 applications (nor can Yale, Princeton, etc.). So they need test scores. The test makers cannot read millions of well thought out answers to complex questions, so they need multiple choice tests. No one needs any of these tests to be about mathematics. (I assume the admissions people don't know algebra or calculus either.)

But, mathematics has a really good property. There are correct answers. Ask applicants what we should do about ISIS or the Ukraine and you can't use multiple choice tests to judge the responses. Someone would have to actually read students' answers. And there would be no "right" answer. 2 +2 really does equal 4. So math wins. And multiple choice tests win. And all our kids lose.

Our kids learn to hate school (because math is boring to most). They lose self-esteem (because they "aren't good at math"). And, what schools teach continues to be irrelevant to the real needs of children.

How about instead of math we teach how to get along with other people? How about teaching personal financial management. Teach kids how to get a job. Teach them to learn real skills (pick any of 1,000). Teach them how to raise a child or how to eat properly. Teach them how to negotiate or how to speak well, or how to plan well.

Ok, enough. Math will win every time for the reasons I stated above and the *New York Times* (undoubtedly populated by editors who majored in English and were "bad at math") will continue to make the country hysterical about why Finland or China have better math scores than we have. I have only one question. Are kids (or adults) happier in those countries?

Silly question. Who cares about that?

---

8 34,295 Apply to Class of 2018, Marking Slight Decrease from Previous Year http://www.thecrimson.com/article/2014/2/3/class-2018-drop-slightly/

Tom Friedman; wrong again, this time about education

I guess it wasn't bad enough that Friedman promoted the Iraq War in his *New York Times* column and then had to admit he was wrong. He actually is supposed to know something about the Middle East.

Now he espousing the nonsense theme of the day, that the problem in education is the teachers. I guess he saw the movie *Waiting for Superman* and wanted to jump on the bandwagon.

So one more time for Tom: the problem is that school is boring and irrelevant and all the kids know it. They know they will never need algebra or trigonometry. They know they will never need to balance chemical equations and they know they won't need random historical myths promoted by the school system. When all this stuff was mandated in 1892 it was for a different time and a different kind of student.

Change the curriculum to something relevant to modern life and you won't need to look for teachers. Teachers will rush to the opportunity to teach kids who actually want to be there.

Don't worry about Artificial Intelligence, Stephen Hawking

The eminent British physicist Stephen Hawking warns that the development of intelligent machines could pose a major threat to humanity.[9]

> "The development of full artificial intelligence (AI) could spell the end of the human race," Hawking told the BBC.

Wow! Really? So, a well-known scientist can say anything he wants about anything without having any actual information about what he is talking about and get worldwide recognition for his views. We live in an amazing time.

Just to set the record straight, let's talk about AI, the reality version not the fantasy one.

Yes, we all know the fantasy one of *2001*, *Star Wars*, or *Her*. We have been watching intelligent machines in the movies for decades.

Apparently, Hawking is using a voice system. That's nice. Maybe he should find out how it works. The new system "… learns how Hawking thinks and suggests words he might want to use next.." So that makes it very smart does it? That is statistics. We can easily count what you have been saying and guess what you will say next. It is not that complicated to do, and it is not AI.

What is AI? AI is the modeling of mind such that you have created a new mind. At least that is what it is to people who don't work in the field. To people who do work in the field, the issue is not what word comes next as much as it how to have an idea about something, or how to have an original thought, or how to have an interaction with someone in which they would think you are very clever and are not a machine.

You average five-year-old is smarter than any computer today and is smarter than any computer is likely to be anytime soon. Why? Because a five-year-old can do the following:

- Figure out what annoys his little sister and do it when his mother is not watching.
- Invent a new game.
- Utter a sentence that he has never uttered before.
- Understand what his parents are telling him.
- Decide not to do it because he has something he would rather do.

---

9 Stephen Hawking warns artificial intelligence could end mankind http://www.bbc.com/news/technology-30290540

- Be left alone in the kitchen and make an attempt to cook something possibly burning down the house but in any case leaving a giant mess.
- Listen to someone say something, draw a conclusion from it, and ask an interesting question about it.
- Find his way to school without help if allowed to do so.
- Throw a ball.
- Get better at throwing a ball by practice.
- Eat certain foods and hate them, and eat others and love them.
- Cry when he is feeling anxious.
- Be thrilled with a new toy.
- Throw a temper tantrum.
- Make his mother think he is the best thing in whole world.

Why am I listing such mundane things as hallmarks of intelligence? Because in order to build an intelligent machine, that machine would have to grow up. It would have to learn about the world by living in it and failing a lot and being helped by its parents. It would have to have goals and tastes and make an effort to satisfy those goals every day. It would not be "born" with goals. I didn't grow up wanting to work in AI for example. That interest developed while I was in college as result of a wide variety of experiences and interactions with others.

If we have to build an intelligence that acquires knowledge and motivation naturally we would have to know how to build the equivalent of an infant and teach it to interact with the world. Would that infant have arms and legs and be trying to learn how to walk and get stuff it liked and be angry and hot and hopeful? If not, it wouldn't be much like a human.

But maybe Hawking doesn't mean AI that is human-like. Maybe he just means a computer program that is really good at prediction by statistics. That is not AI in my view, but it is something. Is it something to fear? Only if you are worried about a machine that predicts certain things in the world better than you can. That could happen.

To build the AI that I have always had in mind requires more money than Mark Zuckerberg is willing to invest and requires a purpose. Before someone builds a general purpose AI they would have to try building a special purpose one, maybe one that is smart enough to kill Bin Laden. Interestingly, while the Defense Department has invested plenty of money in AI it still sent humans to do that job. The Defense Department would undoubtedly have preferred to send an AI robot to do the job, but they are nowhere close to having one.

Could they make one? Yes, someday. But it would be talking to you, or predicting what works, not what Hawking wanted to say next. It would be about navigation and inference and figuring out things just in-time and so on. It would need to know how to talk and comprehend the world – to think really.

Special purpose AI machines, ones that do things like clean our house will be around long before any AI Hawking fears. As much as we all would like one, I don't see any AI cooks and maids around.

The AI problem is very very hard. It requires people who work in AI understanding the nature of knowledge; how conversations work; how to have an original thought; how to predict the actions of others; how to understand why people do what they do; and a few thousand things like that. In case no one has noticed, scientists aren't very good at telling you how all that stuff works in people. And until they can, there will be no machines that can do any of it.

The misreporting of science by The New York Times and others

Reading newspapers about new technology is a lot like going to a fortune teller to find out about the future. Nice stories, but the reality is unknown. Here are the first three paragraphs from a recent *New York Times* article on computers that can give a grade to a college essay:[10]

> Imagine taking a college exam, and, instead of handing in a blue book and getting a grade from a professor a few weeks later, clicking the "send" button when you are done and receiving a grade back instantly, your essay scored by a software program.
>
> And then, instead of being done with that exam, imagine that the system would immediately let you rewrite the test to try to improve your grade.
>
> EdX, the nonprofit enterprise founded by Harvard and the Massachusetts Institute of Technology to offer courses on the Internet, has just introduced such a system and will make its automated software available free on the Web to any institution that wants to use it. The software uses artificial intelligence to grade student essays and short written answers, freeing professors for other tasks.

Sounds great doesn't it? Better service for students, less work for professors, and smart computers – all in one article. Except that it is all nonsense. The *Times* doesn't call the software Artificial Intelligence (AI) but most every other paper printing the same story did. Here is the headline from the *Denver Post* for the same article.

"New artificial-intelligence system grades essays at college level"

We live in a time where every new piece of technology in education is touted as a great breakthrough. Now, AI is my field and my specialty in AI is processing language. No computer can read an essay. Maybe someday, but not now. So, how do they grade them if they can't read them? By counting how many big words were used? By seeing if the sentences are grammatical? The articles that tout the glories of this stuff never actually say how. But no computer can tell if the writer had a good idea. This is actually very hard to determine. It is the reason why well-educated professors would take on the task of reading an essay – to see if there were any good ideas in it. But now that we have MOOCs and everything is about mass education, why bother? No one is listening to anyone's ideas anyway. Just tens of thousands of students hearing the same lecture. Yet, the *Times* and other papers keep touting MOOCs as a great breakthrough. They even had the audacity to mention that professors would now have time free to do other

10 New Test for Computers: Essay Grading at College Level http://www.nytimes.com/2013/04/05/science/new-test-for-computers-grading-essays-at-college-level.html

things. What other things? Their lectures are already recorded and they don't grade papers, so what other things?

The answer, as any professor can tell you, is research. All these MOOCs, essay-grading software, and everything else we are hearing about are meant to allow professors to teach less and do more research. I have no problem with that. I had that view of the world too when I was a professor. The students get shortchanged by this and will get really shortchanged by MOOCs, but neither the *Times* nor the faculty of elite institutions care much about students.

The *Times* doesn't care much about the truth either, at least when it reports about scientific breakthroughs. This is not unique to the *Times* however. The same day, another report hit the press about science:[11]

"Scientists 'read dreams' using brain scans"

This time it was a BBC headline, but many other papers reported the same scientific breakthrough. The scientists quoted in the report did not say anything like this of course. The scientists said that they can now detect images in the brain for some people whom they have studied. The sleeping person is awakened and asked what he was dreaming about and the scientists can detect a similar pattern when it occurs another time. Hardly "reading your dreams."

Newspapers like to make stuff up and people remember the nonsense they read in a headline. So the public thinks that computers can read and understand an essay and the public thinks that the computer can read your dreams. So what if this isn't even close to true? Another paper sold.

One wonders if the scientists aren't complicit in all this nonsense. The answer is yes and no. I am interviewed all the time and I know that the reporter will exaggerate what I said and write a ridiculous headline. I do the interviews anyway on the grounds that some good might come out of it. But many scientists want people to think they are doing stuff that they actually aren't doing. This is particularly true of artificial intelligence, my own field, where the experts quoted in the *Times* article must have known full well how their work would be misinterpreted and didn't care.

Scientists are always selling so that people will get excited and give them more money to do research. And newspapers are always writing headlines that aren't true but catch your eye.

The public loses by being misinformed. At the moment it is being misinformed about education in a serious way. Things in education are not improving. Technology is not helping (although it could). Things in education are getting much worse. Let's see if the *Times* ever says that.

11 http://www.bbc.co.uk/news/science-environment-22031074

I translate Bill Gates dumb remarks on education

Bill Gates recently spoke about the Common Core at a *Politico* event called "Lessons from Leaders," covered on the *Politico* website.[12]

> He cited as a model unnamed Asian countries, which he said have pared down their standards so they give kids "nice, thin textbooks" and focus on teaching fewer concepts in more depth. "And what are the results? Well, they spend far, far less money and get far, far better results," Gates said.

Translation: China has better test scores than the U.S. Are these the results that matter? Then why are Chinese students trying desperately to get into U.S. schools? Is it possible, Bill, that you have no idea what you are talking about?

> He said he thought of Common Core as "a technocratic issue," akin to making sure all states use the same type of electrical outlet.

Translation: Gates wants to plug every kid into a place where they can work as effective parts of the machine. No differences will be tolerated. Curiously, this was exactly the plan in 1900 when the U.S. schools were designed. The goal was to make compliant factory workers.

> "Common Core is, to me, a very basic idea that kids should be taught what they're going to be tested on and that we should have great curriculum material," he said.

Translation: Teaching to the test is now official U.S. doctrine. Teaching is not a discussion but the ramming in of facts. One might ask "Which facts?" That is easy – the ones decided upon by Charles Eliot in 1892, when he was president of Harvard. Great plan, Bill. By the way, facts aren't actually what matters. Clear thinking matters. Oh, except on tests. Bill, did you take a test that allowed you to start a company? No? Then maybe some other skills might be needed in order to learn to do that, eh? Did you learn to do that at Harvard?

In another interview in the *New York Times*, Gates talked about his efforts to change education.[13]

> "We didn't know when the last time was that somebody introduced a new course into high school," Gates told me. "How does one go about it? What did the guy

12 Bill Gates plugs Common Core http://www.politico.com/story/2014/09/bill-gates-common-core-111426

13 So Bill Gates has this idea for a history class http://www.nytimes.com/2014/09/07/magazine/so-bill-gates-has-this-idea-for-a-history-class.html

> who liked biology — who did he call and say, 'Hey, we should have biology in high school?' It was pretty uncharted territory. But it was pretty cool."

Answer: It was 1892. Nothing new has happened since in curriculum creation. Biology is a high school subject because it was a subject at Harvard in 1892. Not a great reason, Bill. Good luck with getting it out of the curriculum. Oh, but you want everyone to memorize the names of phyla.

> Without prompting, he recounted getting a bad grade in an eighth-grade geography course ("They paired me up with a moron, and I realized these people thought I was stupid, and it really pissed me off!") and the only C-plus he ever received, in organic chemistry, at Harvard ("I'm pretty sure. I'd have to double-check my transcript. I think I never ever got a B ever at Harvard. I got a C-plus, and I got A's!").

Response: Everyone should be you Bill, because you are the greatest. I only got C's in college. Maybe that's why I have the perspective that grades don't matter and creativity does. What did Microsoft ever create? Should everyone in the U.S. go to Harvard and get A's Bill? (Except your "moron" friend.) Is that your plan? Or might there be room for other types of people than you, and other kinds of interests than yours or Charles Eliot's?

Back to his speech about the Common Core at the *Politico* event:

> "The idea that what you should know at various grades … should be well structured and you should really insist on kids knowing something so you can build on that. I did not expect that to become a big political issue."

Response: Why not Bill? Here is one explanation. Because you know nothing about education and think everyone is like you and should learn what you learned (which of course had nothing to do with your success). Allowing the possibility that a kid could follow his or her own interests? Nah. Too much like having different kinds of plugs and sockets.

At Cornell University, Bill Gates spoke about the future of higher education.[14] He said that one improvement colleges should make is to cut down on the number of lecturers and focus on the select few who are the best at talking about their fields.

Response: How about having no lectures? Nah. Lectures are wonderful. No one remembers them; everyone is sleeping or texting to their friends. But let's have lectures anyway. They worked so well in the Middle Ages when no one could read. Lectures are about money, Bill. Lots of students in class paying big tuition fees is a revenue model. You don't mention it, but you must love MOOCs too. Soon all the professors can be fired and we can all listen to the best MOOCs.

More from Gates on his vision for measuring the "value" of a university:

14 A Conversation with Bill Gates: Considering the Future of Higher Education http://www.cornell.edu/video/bill-gates-future-of-higher-education

> If you have a high SAT score going in, [the university] is not going to make you dumber. I'd like to see an output measure. I'd like to see a university and say, 'We took kids with a 400 SAT score and they were super smart when they left our university,' not, 'We were sure they were smart when they came in, and we didn't damage them.

Response: Wonderful. More measurement! Your friends would like doing that measurement and making lots of money from it. Why not have college be just a giant test-taking party? Let's eliminate talking with professors who can guide you on a project or seminars where one can learn to express one's point of view and get smarter by arguing. Let all college students simply prepare all day for the exit exam. That will be just wonderful Bill.

Bill Gates: wrong again about rating teachers

An article in the *New York Times* today says that Bill Gates is spending $355 million so that teachers will be rated in a coherent fashion.[15] This rating, of course, has to do with how teachers improve their students' test scores over time. This is an obvious waste of money if you question the value of test scores, which of course, neither the *New York Times* nor Bill Gates have even considered.

For the sake of argument, let's assume that it is important to rate teachers. Here I list **Schank's Criteria** for rating teachers:

- Does the teacher inspire students?
- Does the teacher encourage curiosity?
- Does the teacher help students feel better about themselves?
- Does the teacher encourage the student to explore his or her own interests?
- Does the teacher encourage the student to come up with his or her own explanation for things they don't understand?
- Does the teacher set him or herself up as the ultimate authority?
- Does the teacher encourage failure so that the student can learn from his or her own mistakes?
- Does the teacher care about the students?

Really, how hard is it to recognize that these aspects of teaching are tremendously more important than test score improvement?

15 What Works in the Classroom? Ask the Students http://www.nytimes.com/2010/12/11/education/11education.html

Madrassas, indoctrination, education, and Kristof

It is always disappointing when a writer who says sensible things about most issues decides to turn off his brain when it comes to education. I complained about a nonsensical article about education in the *New York Times* written by Nicholas Kristof a few months ago, and now he has gone and done it again. He is writing about spending less money on troops in Afghanistan and suggests that that money should be spent on education.[16]

> Since 9/11, the United States has spent $15 billion in Pakistan, mostly on military support, and today Pakistan is more unstable than ever. In contrast, Bangladesh, which until 1971 was a part of Pakistan, has focused on education in a way that Pakistan never did. Bangladesh now has more girls in high school than boys. (In contrast, only 3 percent of Pakistani women in the tribal areas are literate.) Those educated Bangladeshi women joined the labor force, laying the foundation for a garment industry and working in civil society groups like BRAC and Grameen Bank. That led to a virtuous spiral of development, jobs, lower birth rates, education and stability. That's one reason Al Qaeda is holed up in Pakistan, not in Bangladesh, and it's a reminder that education can transform societies.

Why am I complaining? This seems reasonable enough. Indeed, Kristof is usually reasonable. And then he says:

> When I travel in Pakistan, I see evidence that one group — Islamic extremists — believes in the transformative power of education. They pay for madrassas that provide free schooling and often free meals for students. They then offer scholarships for the best pupils to study abroad in Wahhabi madrassas before returning to become leaders of their communities. What I don't see on my trips is similar numbers of American-backed schools. It breaks my heart that we don't invest in schools as much as medieval, misogynist extremists.
>
> For roughly the same cost as stationing 40,000 troops in Afghanistan for one year, we could educate the great majority of the 75 million children worldwide who, according to Unicef, are not getting even a primary education. We won't turn them into graduate students, but we can help them achieve literacy. Such a vast global education campaign would reduce poverty, cut birth rates, improve America's image in the world, promote stability and chip away at extremism.
>
> Education isn't a panacea, and no policy in Afghanistan is a sure bet. But all in all, the evidence suggests that education can help foster a virtuous cycle that promotes stability and moderation. So instead of sending 40,000 troops more to Afghanistan, how about opening 40,000 schools?

16 More Schools, Not Troops http://www.nytimes.com/2009/10/29/opinion/29kristof.html?_r=0

On the surface this seems right, but it is very wrong. Americans have the view that Pakistan is full of terrorists and people who take money from the U.S. and make no good use of it. There is some truth to this I assume, but Pakistan is also full of very reasonable and intelligent people who behave a lot like people in the U.S. They go to good schools in Pakistan, they run successful business there, and they worry about fixing their country. I have been to Pakistan a few times, always talking about education and am usually very well received. I have talked with former Pakistani President Mushareff and with various ministers in the government on many occasions. I am on the board of The Beaconhouse School, a private school in Pakistan trying very hard to make great and innovative schooling available around the country.

I have never visited a Madrassa but I have seen the kids that go to Madrassas and they look happy and healthy. Here is a picture I took:

What is the issue here? The issue is indoctrination.

Madrassas have a goal. Their goal is make the kids that attend them believe certain things that the teachers are sure are true and to think and behave in certain ways in their everyday lives. In short, Madrassas, like many other religiously-run schools, know what the end product should be and they have a long history of being successful in creating what they want. The fact that we don't like what they produce is irrelevant.

When Kristof says he wants to build more schools what he means, apart from the obvious – getting kids capable of reading and simple math – is to create more schools like the ones we have in the U.S. In the U.S. we have thousands of schools where kids are packed in like sardines learning a set of subjects will neither help them live their lives reasonably nor help them to make a living. The education they receive is all about getting them into college, which is pretty irrelevant for the majority of the students who just need to be able to function well after graduation.

We offer indoctrination in our schools too. We constantly indoctrinate our children to believe that college is very important and that memorizing facts to help them pass tests is how to get there. This is the system we would be exporting and it is even more useless in Pakistan than it is the U.S. Just saying the magic 'education' word is of no help, Mr. Kristof. You actually have to understand the difference between education and indoctrination. Madrassas do it and the U.S. schools do it too. You are saying that we should indoctrinate Pakistani students with our kind of indoctrination.

I say we should consider what learning is really about and help our children learn things that are, or will be, important to them. (This would not include say, our indoctrination about the significance of algebra or the wonderfulness of our glorious history.) Build a school that does that, use it to help our own children learn, and then export that.

The U.S. schools aren't as good as Madrassas. They have no goal, they don't know what they want to produce, and they have no agenda at all except raising test scores. How would spending millions on building these kinds of schools, the ones with the horrific drop-out rates, and the pregnant students, and the drug dealers on campus, be a good thing? We are not doing so well over here in education.

The Beaconhouse School in Pakistan is every bit as innovative as any school we have in the U.S. Here is a picture of me helping the teachers at Beaconhouse think about learning:

More absurdity from the New York Times and Nick Kristof

I must say I am getting used to reading ridiculous columns about education on the *New York Times* op-ed page, but yesterday Kristof's took the cake.[17] He made up a paragraph to test reader's knowledge of the Bible that was intentionally full of nonsense. It started like this:

> Noah of Arc and his wife, Joan, build a boat to survive a great flood. Moses climbs Mount Cyanide and receives 10 enumerated commandments;

He then asserts that most Americans wouldn't be able to spot the errors he deliberately made, which might well be true. And then he says:

> All this goes to the larger question of the relevance of the humanities. Literature, philosophy and the arts have come to be seen as effete and irrelevant, but if we want to understand the world around us and think deeply about it, it helps to have exposure to Shakespeare and Kant, Mozart and Confucius — and, yes, Jesus, Moses and the Prophet Muhammad.

So because Americans are generally ignorant and think Joan of Arc was married to Noah we should teach them Kant, Mozart, and Shakespeare. Wow!

First I'd like to point out to Mr. Kristof that we do teach Shakespeare, and most Americans get a lot of information about Jesus. Many of the Americans who would not spot the errors Kristof made have indeed learned about Mozart.

And that is the point. All that teaching didn't work to do what Kristof imagines school does. Students don't remember what they are taught for the most part. Yes, the more intellectual students remember some of what they learn, but the more intellectual students are not confused about Joan of Arc. If the people who have trouble with Mr. Kristof's paragraph were taught to think more clearly, it might help. But that is not the point.

The issue here hasn't changed for years. We are told, "Americans don't know where the Ukraine is and if we only told them more about the Ukraine we would all be better off." This is a tragedy, but it is not the concern about what Americans know that is the tragedy, and it will not be fixed by more humanities study.

The real tragedy is that most students find school so unappealing that they just work to pass the course and immediately forget everything they learned. The second tragedy is

17 Religion for $1,000, Alex http://www.nytimes.com/2014/04/27/opinion/sunday/kristof-religion-for-1000-alex.html

that even if they remember what they learned – how to balance a chemical equation, what Hamlet said about Yorick, or what Beethoven's Fifth sounded like – it would make no difference. This information does not make them better thinkers, more capable of earning a living, more capable of making well thought out voting decisions, or more capable of having better and more meaningful relationships with other people.

Having more of the Great Books crammed down one's throat just makes one resentful of the cramming. Not everyone wants to be an intellectual or is likely to be an intellectual, something that curriculum designers tend to ignore.

I am glad Kristof wants to learn about all the religions in world. He should do that. But the four he refers to are just four of many. Why not learn about every religion?

If a student likes Shakespeare, let him read (or better yet see) a play or two. We simply have to stop this idea that we must decide what everyone needs to know and then ram it down their throats. Students have been resisting that ramming for as long as there have been schools, which is exactly why Kristof is correct in assuming grown-ups don't know much.

School has to change radically, but the *Times* keeps wanting more of the same. The subjects we teach were decided in the Middle Ages. A lot has happened since then. I don't notice Kristof demanding more computer science, or more psychology, or more biotech. Those subjects help people think too, you know. He promotes the humanities because that is what he knows, I assume. There is nothing wrong about learning from the humanities, but the argument for it would have to be stronger than that made in Kristof's column.

The fake choice: preschool or prison

Nicholas Kristof, the *New York Times* writer who writes interesting columns about human rights issues around the world, every now and then veers off into talking about education, a subject he clearly doesn't understand. Maybe he thinks he understands it because he went to Harvard, or maybe the *Times* is telling him to push their schooling agenda. I don't know. What I do know is this week he wrote, "Do We Invest in Preschools or Prisons?"[18]

Really? Is that our choice? Lately the Obama administration has been pushing pre-kindergarten as something the government should invest in and force upon all children. Just what kids need – more school. The idea that the choice is preschool or prison is an interesting one. It comes from the fact that people in prisons are disproportionally undereducated, poor, and not likely to be white. While all of this may be true, pre-kindergarten would not solve the problem.

As it happens, my granddaughter is in pre-K right now. The other day she recited the planets to me. She had no idea what a planet was, but she knew their names. Well, that will certainly keep her out of prison. Oh, wait. She wasn't going to be going to prison because she has caring, educated parents and she is not poor.

How is pre-K supposed to fix the very serous problem of the giant underclass that we have in this country? One way to understand this issue is to listen to the parents of these underclass children speak. They sound as you would expect, not educated and incapable of speaking English clearly or forming coherent thoughts. But would they speak better English and have clearer thinking if we made them go to school more?

Actually, adults who have these at-risk children (who need pre-K) did go to school. Typically they are high school graduates who still can't speak well, can't think clearly, and they cannot parent well either. Probably they weren't parented well themselves. School did not fix this.

Why is that?

It is because the high school curriculum they took is absurd. The issue is not teaching kids the names of the planets when they are four nor is it teaching them about amoebas or George Washington when they are fifteen. The issue is teaching teenagers how to manage their own lives. Could we teach them how to be parents? Could we teach them

18 http://www.nytimes.com/2013/10/27/opinion/sunday/kristof-do-we-invest-in-preschools-or-prisons.html

how to make a living? Could we teach them how to have better human relationships? No, clearly we cannot, because that is not what high school teaches.

Mr. Obama and Mr. Kristof don't seem to understand this because there is a school lobby run by Bill Gates and the testing companies that has effectively taken over education. That lobby wants to sell more school, more tests, and less freedom to determine what it is you may want to learn about.

Let's imagine for a moment that instead of teaching every teenager algebra we taught them how to reason clearly about their own lives. Let's imagine that instead of teaching them English Literature and asking them to write papers about what Dickens' main themes were, we taught them to speak well, diagnose their problems, and write a clear plan about how to accomplish something they want to accomplish. Let's imagine that instead of teaching them to memorize facts about plants or planets we taught them to take care of their own health needs or taught them to reason from evidence.

Make better parents and you won't have to shove pre-K down everyone's throats. Keep teaching more math and science and you will need more prisons. Why? Because only a small percentage of students will care about, or will succeed at, math and science. Teach parenting and job skills in high school, and you will need fewer prisons.

Measuring teachers means education reform? You have got to be kidding!

Last week, in the *New York Times*, there was an op-ed column contributed by a Professor Emeritus (of Nursing) from the University of Maryland.[19] Why the *Times* considers this man's opinion worth publishing is anyone's guess, but his article fits in well with the *Times*' continuing insistence on always being on the wrong side in education.

The article starts with this gem:

> Of all the goals of the education reform movement, none is more elusive than developing an objective method to assess teachers.

Really? That is the issue? Measuring teachers? Funny. I thought the issue was making schools that excited students and made them into people who loved learning and were learning things that they chose to learn and were excited to learn. Silly me.

I was a pretty good teacher if I do say so myself (and many of my students say exactly that in my book, *Teaching Minds: How Cognitive Science Can Save Our Schools*). But I couldn't make algebra interesting to those who are bored to death by it. And, I couldn't make literature interesting to those who think reading nineteenth century novels is tedious and irrelevant. In fact, I avoided teaching introductory programming my entire career because there was no way that I could make that interesting. Now, there are people who can make these subjects interesting. (Saul Morson and Chris Riesbeck, both at Northwestern do exactly that in their respective subjects.) But they have an advantage. No one makes students at Northwestern take Russian Literature and no one makes them takes Introductory Programming either. Motivation matters.

But this is not the case for the high school teachers that this nursing professor wants to measure. (One would assume nursing students take nursing because they want to be nurses by the way, which would have made his job as a teacher a lot easier.)

No, he wants to measure, "...the amount of time a teacher spends delivering relevant instruction."

Really? This sentence is so wrong on so many levels that I find it impossible to believe this man was ever a teacher.

19 A New Measure for Classroom Quality http://www.nytimes.com/2011/05/01/opinion/01bausell.html

Let's start with the concept that the job of a teacher is information delivery. This model of teaching is not only out of date, it is simply wrong. If it were right, you could apply the speed principle. If one teacher were to talk twice as fast as another teacher, he or she would deliver twice as much information and thus be twice as good.

A teacher's job in today's world is unfortunately to get students to do well on standardized tests that test how much information you can temporally memorize and how many test-taking tricks you know.

Here is another gem from this article: "…the teachers who taught more were also the teachers who produced students who performed well on standardized tests."

Wow! Teaching couldn't possibly be about motivating students or helping students be better people or helping students think well or live their lives well. No, it means teaching more – really, teaching faster would do the trick – and not even noticing if anyone is listening or if anyone even gives a hoot about what you are teaching. Test scores! Test scores! Test scores!

What about re-thinking the subject matter that we teach and the idea that classrooms are really bad places to learn?

The *New York Times* has never had a clue about education, as I have said many times before, but this article is a new low. As one Emeritus Professor to another, I suggest that Mr. Nursing Professor go back to thinking about how to teach nurses and leave education reform to those who have some idea what the real issues are.

Teachers are not and have never been the problem. You can't make algebra interesting to someone who isn't interested in it. Teachers are forced to rely on that old canard "you will need it later" which is, of course, simply untrue.

Fared Zakaria and Ivy League graduates keep defending the liberal arts, but clearly the liberal arts didn't teach them to think

The problem with working on changing education is that everyone has an opinion. You went to school didn't you? So, you are an expert. And, if you have a well-known name because you were on TV a lot about entirely different issues, you are still an expert on education. Fareed Zakaria has published a book on the value of a liberal education and a part of that book appeared in the Washington Post.[20] Here is a quote from that article to give you an idea about his point of view:

> From President Obama on down, public officials have cautioned against pursuing degrees like art history, which are seen as expensive luxuries in today's world. Republicans want to go several steps further and defund these kinds of majors. "Is it a vital interest of the state to have more anthropologists?" asked Florida's Gov. Rick Scott. "I don't think so." America's last bipartisan cause is this: A liberal education is irrelevant, and technical training is the new path forward.

Since I am always talking about education, I can tell you that this is a very typical response to what I say. Earlier this week I was asked about Shakespeare: *Don't you think kids should still read Romeo and Juliet?* And in a different conversation the same day, when I questioned the wisdom of teaching algebra: *But algebra teaches you how to think.* My usual reply is that it is sad that these people never were able to think before they learned algebra.

The issue is neither liberal arts nor algebra nor the idea of training everyone to become a programmer. I think people should learn what they want to learn. What a radical idea! Teachers should be guides and mentors, not fountains of knowledge. Learning should be fun. We should not "teach evolution" nor should we *not* teach evolution. We should not teach Dante or Cervantes (which any Italian or Spaniard will tell you we **must** teach). We should let kids follow their own interests.

What are their interests by the way?

20 Why America's obsession with STEM education is dangerous https://www.washingtonpost.com/opinions/why-stem-wont-make-us-successful/2015/03/26/5f4604f2-d2a5-11e4-ab77-9646eea6a4c7_story.html

The Bureau of Labor Statistics polled several hundred children who live in and around New York City (in 2012) who were between 5 and 12 years old. These are the career aspirations that they found the kids had in order of most desired:

1. Astronaut
2. Musician
3. Actor
4. Dancer
5. Teacher
6. Firefighter
7. Policeman
8. Writer
9. Detective
10. Athlete

In the U.K. they surveyed 1,000 children aged 6-16 and the results were similar. They found that the top ten dream careers for children were:

1. Professional Athlete
2. Performer
3. Secret Agent
4. Fire fighter
5. Astronaut
6. Veterinarian
7. Doctor
8. Teacher
9. Pilot
10. Zoo Keeper

Since it was the STEM Centre that did this survey they determined that were all STEM careers and wasn't that wonderful?

Could we just let kids be firemen (in simulation) until they get bored with that and then let them keep a simulated zoo? Could we let them try to be detectives and astronauts (in simulated worlds) or let them try to be actual writers and actors if that is what they want to be?

Why wouldn't it be the school's job to make sure that the fireman curricula taught about the physics of firefighting and the chemistry of what causes fires and how to deal with stressed people and how to address the public in a crisis? Are these things STEM or are they the liberal arts? Who cares?

Could we let an aspiring actor play Romeo but allow him to research the part and think about how to rewrite Romeo for modern times and to learn why Verona was different from modern day Duluth? Why can't we help our aspiring musicians learn to think hard about music, write music, and figure out how the music business works? Could all those things teach you to think too?

What definitely does not teach you to think is learning the right answer to put on a multiple choice test about Romeo and Juliet, or Cervantes, or Dante.

The problem here is that any university graduate (especially ones from the Ivies it seems) think that the courses they were forced to take in college have broadened them and made them better people. (This is the very definition of Cognitive Dissonance.) They never got to live an alternative life however. They never got to do something other than sit in a classroom and listen to lectures and prepare for tests or write essays about subjects they were forced to study but may not have found very interesting. People are different. They should be allowed to be different.

Our idea of education is a very elitist one. We are worried that everyone should have to read Romeo and Juliet. But why? How often does that come up in real life? We don't need literature in order to discuss these same life issues. Discussing life through the works of Shakespeare sounds appealing to intellectuals, but it really is the hard way to do it and may never actually get the attention of most of the population who could be, and should be, in these same discussions.

I, on the other hand, am worried that everyone should be capable of asking hard questions of politicians who spout nonsense and that everyone should learn to do something that is valuable in the society in which they live so they can earn a living. By no means do I think that they should be taught only technical skills but neither do I think that kids should be forced to study the liberal arts.

We learn to think by thinking. We think even as small children, amazingly, without the help of algebra or art history. What happens is that people stop kids from thinking by telling them "the truth" and failing to have conversations with them that might challenge their beliefs or force them to defend their ideas. We learn to think through intellectual engagement and intellectual combat, not through indoctrination.

Our entire notion of school is wrong. We need to stop "teaching" and we need to start letting kids explore their own interests with adult guidance. There is no need to defend the liberal arts. Make the choices interesting and then give them many choices. By this I do not mean choices of courses to take. Enough with courses and classes. Let them choose experiences to have. It is our job to build potential experiences for them, guide them through the ones they have chosen, and offer alternatives when they change their minds.

Chapter 9: Milo

My grandson visits

My grandson (Milo) visited for a few days this week. He is just shy of one and a half. Here is some of what I learned from his visit.

The toys they make for kids are just awful. He had a drum made by LeapFrog that did everything but make drum noises. It was constantly yabbering about what he had just learned when he hadn't learned anything. He liked turning it on and off.

There is a Barbie computer that makes the LeapFrog drum look like a work of genius. It was made for girls but I thought he'd like it because his parents are always keeping him off their computer. He did like it. But it made no sense. It actually asks the kid which friends he doesn't want to talk to. He just wanted to push buttons and make something happen. I am sure the Barbie geniuses will say it wasn't made for his age. Maybe so. But I wouldn't let any girl for whom it was made near it.

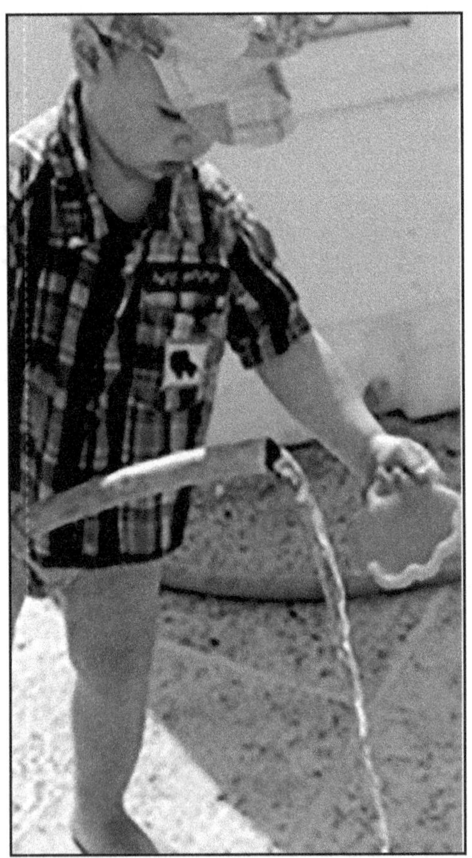

Milo learns about hoses - no toys or books in sight

The books they have for kids don't teach kids much of use.

My complaint is not that the stories are absurd, which they are, nor that no kid could really understand them, which they can't, nor that the stories are so awful that the real lesson they teach is that they should be kept away from kids, which they should be.

No, my complaint is that they are clearly meant to entertain adults. Kids should be reading about stuff that actually happens to kids or that actually happens to the adults in their lives. Stories should be about going to the bank, or the park, or watching food being made, or driving in a car. Stories should not have plots. Plots are for adults.

Kids need to figure out how the world works and we confuse them with literary devices meant to amuse the parents.

Of course the real issue here is that this state of affairs continues on in school. Schools are meant to satisfy parents (and politicians). So we teach stuff that doesn't matter to kids and impresses adults – like how to find a country on a map or how to find lines of symmetry.

I have an idea. Why not meet kids where they are and ignore what adults want.

Nah, too crazy. It would result in kids who actually wanted to go to school and we would all be confused by that.

A drum should be a drum

My grandson Milo visited again. He is nearly two. His visit started me thinking about the toys I had bought him for the last visit and what they tell us about the current view of education and parenting that in our society.

Simply put, most of the toys I bought him are absurdly stupid. He loves a truck that he calls the digger because it is just a big toy truck. But he doesn't much care for the drum, the computer, or the phone that I bought him. He does like the house's hose a lot, as well as the water guns and the sand toys.

I am with him.

The drum, the computer, and the phone suck. It is obvious why. They were designed for parents not for kids. They are meant to teach instead of to be played with. But what they actually teach and what they purport to teach are different things. The real question is "What the heck is going on here?"

With this question in mind I visited the web sites of three toy companies to see what they think they are doing. I started with LeapFrog.

> Brightlings Builders hold fun surprises that encourage exploration! Nine pieces fit together in different ways, motivating babies to create endless combinations. Different shapes and sizes develop spatial awareness. Colors, textures and sounds stimulate all the senses. It teaches Motor Skills and Creativity.

Until I read this, I wasn't really sure how to teach creativity, a subject I actually do know something about, but now I know how. You get a LeapFrog toy with nine pieces that fit together in different ways and voilà! Do you think LeapFrog has done the long-term study that shows that babies who use this toy get more Nobel Prizes? I guess they haven't had the time.

Here's another from LeapFrog:

> The Learn & Groove™ Activity Station introduces infants to learning through music and play. A flip of the jukebox page transforms the activities on the table from learning activities into musical discoveries. In Learn mode, activities on the table help develop fine and gross motor skills while teaching infants the alphabet, number names, counting 1-5, colors, language development, and cause & effect.

Wow! Kids learn cause and effect because they push buttons and something happens! Don't they learn this from falling down and throwing food as well? And gross motor skills! Hoo hah. I thought you learned those by moving around. Well, what do I know?

One thing I was beginning to see from reading these ads is that there must be a lot of parents out there worried that their kids will grow up color blind. I never met anyone who couldn't recognize and name colors. Maybe it is because everyone has had these wonderful teaching toys!

Maybe it is just LeapFrog that is insane. They were the ones who made the drum I bought for Milo that names letters when hit instead of making the drums sounds I had expected. So I looked at some others companies' sites.

> Bring out the genius in your pre-schooler by enriching his play time. We combine colorful characters and advanced technologies to get a child's attention. But, as a parent, you'll love the fact that he's getting ready for school by learning letters, numbers, writing and counting. Entertain your preschooler and a lifetime of learning will begin. Your little one will be 'wired for learning' with Baby's Learning Laptop! The colorful keyboard interacts with a bright light-up screen to teach shapes, common objects and feelings.

And suddenly I was beginning to get it. These companies (this time it was VTech) are catering to a game that parents think they must be playing. They are bringing out the child's genius because all children are potential geniuses. They just don't make average kids any more. The non-geniuses are those are that didn't get "genius-building" toy computers.

Your kid can't just go to school one fine day when he is five. He won't succeed without preschool. But wait. What if he isn't ready for preschool? He must be "wired for learning" and he must be taught shapes, because the names of shapes will get him ready for geometry which will get him the high SAT scores he will need to get into Harvard. So make sure he starts practicing for those standardized tests when he is two.

Whoops. I forgot it also teaches feelings. He is going to learn feelings from a toy computer! At least they have that right. He will learn frustration, irritation, and boredom. Can't start too early on those. But wait. It gets worse.

> The V.Smile™ Baby Infant Development System goes beyond passive developmental videos with a breakthrough, interactive approach to learning. Select the Play Time mode on the panel and watch your baby play and use colorful, easy-to-press buttons to hear fun, educational phrases. Select the Watch & Learn mode and your baby can watch educational animations complete with baby sign language. Finally, as your baby grows, select the Learn & Explore mode where he can actually direct the play on the screen by choosing the subjects he wants to explore. Each Baby Smartridge features learning games with five different baby signs, and teach important skills like colors, numbers, sounds, animals, music and shapes.

In other words – it's a TV. And your baby will learn even more colors and shapes from TV. Plus there are "educational phrases!" Oowee! And it's "interactive"! Just like TV! There are those interactive on, off, and channel-changing buttons and everything.

Do people really buy this junk? Yes, they do. Even I did. But I didn't know what I was getting. This is what is available. What else is out there? It is hard to find a drum that is just a drum.

Finally there is Baby Einstein. I even hate the name of that company. Here are two of their products:

> Baby Einstein Eat & Discover 9-oz. Insulated Straw Colors Cup. Baby Einstein makes mealtime easy for busy toddlers. From the non-spill top with a fun straw to the secure bottom, he gets to take charge of feeding himself – just the way he likes it! Makes mealtime an opportunity to discover colors, shapes and nature.

Since learning about food and how to eat it is so unimportant, let's distract your possibly colorblind child with more nonsense. No time for a break from all those educational toys. Now even the cutlery is educational.

This next one is my favorite however:

> The Baby Einstein Color Kaleidoscope introduces your little one to color in three languages.
>
> Mode One (3-6 Months) introduces your little one to primary colors. Grasping the brightly colored handles, baby activates color-coordinated lights in red, yellow and blue accompanied by enchanting melodies for an engaging light show.
>
> Mode Two (6-12 Months) introduces your little one to three languages. Grasping either of the red, blue or yellow handles, babies will hear the name of each color in English, Spanish and French.
>
> Mode Three (12+ Months) introduces baby to secondary colors. By grasping more than one of the multi-colored handles, your little one will mix the primary colors into three secondary colors of green, orange and purple for an entertaining show of light and music.

In addition to the continuing color obsession, now we have some new idiocy. While your child is trying to sort out the language that his parents speak to him, let's add some stuff that even his parents won't recognize. Because, it has certainly been shown that children grow up trilingual if only you remember to say some words of another language to them from time to time. Colors are so important for mental development because it has been shown that people who are colorblind are also incapable of functioning in society especially if they don't know how to say the names of those colors in three languages.

All right. Enough sarcasm. What is going on here?

Clearly toy companies are trying to cash in on two things. One is the fear that your child will lose the competition and not get into Harvard and the other is that the schools are incapable of teaching the basics.

This is all very sad. Not that many people actually get into Harvard, after all, nor is Harvard so darn important. I have actually known successful, happy, people who have not gone to Harvard. (OK, I used to teach at Yale, I admit it.)

And, the schools are capable of teaching the basics. Your child does not need to go to preschool unless you need the day care. And he does not need these silly toys at all. A child needs a safe environment, loving parents, and a world that stimulates him and allows him to practice skills he needs. What skills are those? Talking. Listening. Navigating the world around him. Seeing what the stuff in his world does. Seeing what the adults in his world do and trying it out.

I had hoped the phone I bought Milo would let him pretend to talk on the phone, but the phone makes so much racket that you can't pretend anything. Real phones don't say letters when you hit their keys. That's not what they are for. Toy phones are for fantasy play about talking to someone, I bought him the computer so he could pretend to send e-mails and copy what his mother does. But instead he hears about pretending to download a screensaver and excluding other kids from the conversation. Fortunately he doesn't understand any of this. When he does I will throw the damn thing out.

The schools, I am afraid, have achieved in the public's eye, the status of a place that is so broken that your kid is better off learning from toys. Now, I do think the schools are broken, but parents are broken too. (This is one of the reasons why the schools are broken.)

Parents: Stop buying educational toys. Sit down and talk to your child and teach him to do what you do by letting him try it out when he is ready. "No," to interactive toys. "Yes," to interactive parents.

By the way, Milo knows the alphabet. He didn't learn it from a drum.

Milo goes to kindergarten. We need to fix this fast.

Milo started kindergarten last week. As it happens I was in Brooklyn this week, so I asked Milo if he liked school. He said he did. I asked him what he liked about it. He said he liked recess. I asked him if there was anything else he liked. He said, yes, he liked lunch. Anything else? "Snack time." Anything else? "Choice time" (Apparently you can do whatever you want then.) At this point his mother said, "What about science?" (His class has been learning science words or something like that.) He said, "No, that was boring."

So it took one week for Milo to learn to say the most common word used to describe school by students: *boring*. (This tidbit of information I owe to my friend Steve Wyckoff, former superintendent of schools in Wichita, Kansas and now an education reformer.)

Milo had a homework assignment. His mother called to ask me what to do about it, because he was already refusing to do it. He was to circle all the "t's" in some document. Since Milo can already read (and write in his own special spelling) he also found this assignment boring. I told her to explain to the teacher that Milo wasn't going to do things that seemed irrelevant for him to do.

All of this made me start to invest more in our Alternative Learning Place idea, opening in Park Slope, Brooklyn, in September 2011. If Milo has to endure the New York City Public Schools for more than this year of kindergarten, I am sure that we all will be driven to drink.

Milo and the Park Slope parents meet Common Core; everybody loses (except Bill Gates and Big Pharma)

Milo has finally met the Common Core, that brilliant new basis of our education system that will kill traditional schooling better than anything I could propose.

Take a look at the math vocabulary words Milo and his third grade comrades had to memorize last week.

rectilinear figure	number line	array
rhombus	Identity Property of Addition	Associative Property of Addition
square unit	place value	customary system
standard form	area model	Distributive Property
unit fraction	arithmetic patterns	endpoint
expanded form	fact family	line plot
expression	Order of Operations	

Before I start getting upset here, let me just point out that vocabulary tests are not mathematics education. Now for the part where I get upset.

My daughter sent this list to me because the Park Slope Parents group (this is an upscale part of Brooklyn, so parents are generally well-off and well-educated) was extremely upset by all this. It seems very few of the parents knew any of these mathematical terms. Now, I have to admit, I do know them (not sure what "expanded form" is though) because I majored in mathematics in college. That is where this stuff comes up. Even in that context, I never found any of it useful in my actual life.

So, parents in Park Slope are busy learning math vocabulary to help their kids who have been confounded by Common Core. What the point of all this is, I am not sure. But here is something I am sure of. There is big money behind all this.

Today the *New York Times Magazine* printed the following article, "The Not-So-Hidden Cause Behind the A.D.H.D. Epidemic."[1]

I tweeted that article to my followers and one of them (who is the head of a Day School in New York) tweeted her favorite line from that article, "A.D.H.D. diagnosis increased by 22 percent in the first four years after NCLB was implemented."

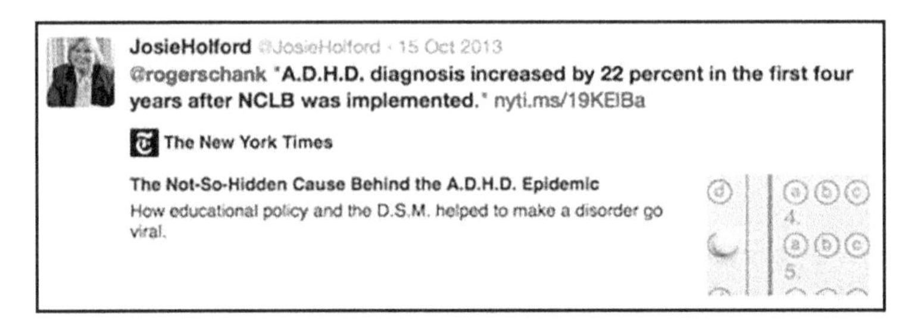

A connection between ADHD (Attention-deficit/hyperactivity disorder) and school testing is hardly a surprise for those of us who have been following the ADHD story for years now. ADHD is a convenient diagnosis because it sells a lot of Ritalin and makes Big Pharma happy; it quiets down unruly kids so the school is happy; and, it makes everyone much more focused on testing so the big testing companies are happy. More Ritalin, better test prep sales.

Common Core will surely help everyone make more money. But poor Milo. Milo suffers, big business wins.

Who cares about any of this mathematics anyway? Who needs it?

Here are two homework questions from Common Core that he had to do last week (among many others):

Homework

1. Fred has 10 pears. He puts 2 pears in each basket.
   a. Draw an array where each column represents a basket of pears.
   b. Redraw the pears in each basket as a unit in the tape diagram. Label the diagram with known and unknown information from the problem.
2. Ms Meyer organizes 15 clipboards equally into 3 boxes. How many clipboards are in each box? Model the problem with both an array and a labeled tape diagram. Show each column as the number of clipboards in each box.

My daughter said that parents were watching YouTube videos in order to figure out what any of this meant. Of course, it is now the parents who are doing the homework (as well as teaching mathematics that they themselves do not understand).

1 http://www.nytimes.com/2013/10/20/magazine/the-not-so-hidden-cause-behind-the-adhd-epidemic.html

Just in time for all this I read a good interview with a mathematician from UC Berkeley.[2]

This is his answer to the question, "*What is wrong with the way most of us are introduced to math?*"

> The way mathematics is taught is akin to an art class in which students are only taught how to paint a fence and are never shown the paintings of the great masters. When, later on in life, the subject of mathematics comes up, most people wave their hands and say, 'Oh no, I don't want to hear about this, I was so bad at math.' What they are really saying is, 'I was bad at painting the fence.'

So, Bill Gates, you have succeeded. At what or why you did it I do not know. But you will make every child and their parents miserable. You will create students who hate school (even more than they do now).

Bill and I, as it happens, have a mutual friend. My friend asked me if I wanted to meet Bill. I said, "Would he listen?"

My friend said "No."

My advice to parents: Fight back. Refuse to drug your kids and refuse to have them take Common Core tests. As they used to say in the old days: "In Union there is Strength."

2 What Is It Like to Be a Mathematician? http://www.slate.com/articles/health_and_science/new_scientist/2013/10/edward_frenkel_on_love_and_math_what_is_it_like_to_be_a_mathematician.html

Milo's mom fights the Common Core

Milo's mom, my daughter, is getting upset now; she wrote this column. For some background on her – she has written a book, published many articles, and designs web sites.[3]

Breaking News: 41-Year-Old Still Bad At 3rd Grade Reading Comprehension
Every night when Milo asks for help with his homework I get a little nervous. As long as he needs help with a researching a report or writing something or practicing spelling words I feel on safe ground - these are things I know how to do – but I've been waiting for the day when I have to tell him I simply don't know the answer to something. That day turned out to be yesterday.

Milo reported that his class had taken a practice "assessment" (they don't call them tests because they don't want the kids to freak out about being tested all the time, so they call them assessments so that instead the kids can eventually freak out about being assessed all the time). He had gotten two answers wrong on the assessment so his assignment was to change the answers to the correct ones and explain why those were the correct answers. Since he wasn't sent home with an answer sheet this meant his homework was really to guess at the correct answer, make sure I agreed, and then come up with an explanation for the answer we'd agreed upon.

Much to my dismay, the assessment he brought me wasn't math homework, which I still feel pretty confident with since we're at a 3rd grade level, but reading comprehension. I totally, totally suck at reading comprehension. Or, more accurately, I suck at reading comprehension "assessments." I scored the same on the math and verbal SATs despite the fact that I never really got math and spent huge quantities of my childhood with my nose in a book.

I took Milo's reading comprehension assessment and sighed. This was going to be okay now, right? After all, I'm an adult. I read lots of books and one assumes I comprehend them or I would have stopped reading long ago. Not only that, Simon and Schuster and the *New York Times* agree that I'm a writer. No published author could be bad at reading comprehension, right?

The first thing I did was look over the questions and the answers, which was how I always approached reading comprehension as a kid. The passages they give you to read are always so mind-numbingly boring that my usual strategy was to see if I could answer the questions without actually reading the passage (wait, maybe that explains why I never did well on these things ... but I digress). My heart sank as I realized in

3 https://uxmag.com/contributors/hana-schank

order to help Milo with his assessment I was going to have to actually read the passage. It turned out to be a mind-numbingly boring passage about a kid who went to camp to learn to swim. He didn't want to be in the group with the non-swimmers, even though he couldn't swim, so he kept asking when he could be moved up into the group for swimmers. Day after day he goes to the camp, does the stuff he's supposed to do, and asks if he can be moved up to the group with the kids who can swim. Eventually he learns to swim and gets moved up into the group. The end. ARE YOU STILL AWAKE?? Just checking.

So the first question Milo had gotten wrong was something like:

The kid in this story is:

a. lazy
b. keen
c. reckless
d. angry

Milo had put down that the kid was lazy.

"I get that," I said to him. "I totally get that. You put down that he's lazy because he didn't want to have to do all the things he had to do to learn how to swim, right?"

"Yes," said Milo. "He just wanted to go right to the group for kids who knew how to swim."

"Here's a tip that it took me a really long time to learn," I said. "The answer is never that the main character is lazy. Or mean, or evil, or a slob. The main character in these things is always something nicer than that. I can't explain why, but that's how they write them. Even though you're right. He is kind of lazy."

"So maybe it's reckless?" Milo said.

"Maybe," I said. "I mean, anyone who doesn't actually take the time to learn to swim and just tries to swim is a little reckless. It's not angry, though I could make an argument for why it could be angry. Maybe he's angry about having to be in the group for non-swimmers."

"Is it angry?" Milo asked.

"It's not angry," I said. "Let's look up what 'keen' means." Milo was shocked that I didn't know what it meant. I explained it's a word that no one has used in the last fifty years, so it makes perfect sense that it would appear in a reading comprehension assessment for eight-year-olds. The definition for "keen" is 'confident'. The answer was "keen."

We moved on to the next question. And for the life of me I couldn't figure out the answer. I found myself making an argument for every single answer. They all seemed equally valid. And then I remembered why I couldn't do reading comprehension as a kid. No matter how hard I tried I couldn't stop arguing with the people who wrote the test. I always felt like if they would just give me the chance to make my case in person I could convince them to see it my way. I wanted to accompany my responses with long essays about how all answers could be right if viewed in the right way. It wasn't that I didn't care. It was that on some level I cared too much about writing and reading to fill in a letter on a bubble sheet and move on. I always found myself editing the passages as I read them, thinking about how I would rewrite them. As I read through the answers I saw them all as correct because writing is fluid and open to interpretation and that is what makes it such a joy to experience. One person may see a kid in a story as lazy and another may see him as angry and they are both right, and if you don't understand that then you haven't comprehended anything about what you've read.

In the end I had to ask Steven what the right answer was. He knew. He always knows what the people who wrote the test are really asking for; it's a skill I will apparently never acquire.

When I was a kid I took my failure at reading comprehension personally, as though the test takers were telling me that I'd never be a writer, that my love for reading was misguided, that the one thing I thought I was good at was a lie. I don't take it that way any more, but there is still a small part of me that wants to print out the list of places I've been published and mail it to the makers of this assessment, along with a clearly thought out essay on why "lazy" could be the right answer. Because really, it could. Take another look, Common Core people. Open your minds a little. You might find the main character a little lazy.

Milo takes a computer class and eventually will learn PowerPoint. Yippee!

My daughter, Milo's mom, sent me a report on a conversation she had with Milo last week.

> My Daughter: I sent Papa a list of all the stuff you're going to learn in your computer class up through 5th grade.
>
> Milo: I bet he hated it.

Right Milo.

I hated it.

Here for example, is the 4th grade computer class (which as it happens is more of less the same as the 3rd and 5th grade computer class):

The 4th grade year will look like this:

- Review log-in procedures
- Microsoft Word
- The basics - opening and quitting
- File management & navigation - Save As + Open: saving a file properly (proper name, location, etc) and finding the file and opening it up on a different occasion
- Basic typing review - proper way to capitalize, using "Shift" for certain punctuation, 1 space between words, etc.
- Practice typing - copying a document
- Formatting 1 - margins, when to use the return key, etc.
- Formatting 2 - Bold, Italic, Underline, etc.
- Formatting 3 - changing font size & style
- Practice formatting - using typed document & changing font style, size, bold, etc.
- The Internet
- The basics of Firefox/Chrome - the buttons, address bar, tools, etc.
- Searching using Google - what the results mean, pictures, etc.
- Downloading pictures - proper naming and saving procedures
- Combining Microsoft Word and the Internet

- Inserting pictures found on the Internet into a Word document
- Formatting the picture in Word - size, position, etc.
- Formatting the complete document - text and pictures
- Microsoft PowerPoint research project
- Basics of PowerPoint - how to create, format, etc.
- Creating a practice Presentation with text, pictures, animations, etc
- Using the Internet for research - proper searching techniques, identifying reputable sites, etc.
- Taking the information gathered on the Internet and creating a PowerPoint Presentation conveying what they have learned - text, pictures, animations, etc.

Really?

This is computer education in the New York City Schools? We could teach kids to program you know. Or we could teach them to build apps. Or to create art. Or to build robots. Or to create a web site. Or to create music. The list could go on and on.

But, you know what Milo learned yesterday? How to create a hashtag.

Great.

Twitter is now part of the curriculum. As far as I can tell kids have been able to learn to tweet all by themselves (for better or for worse).

So, we can't modernize the curriculum because of Common Core and when we can do something small, like add a computer class, it is to teach stuff like PowerPoint.

School is getting worse all the time.

Why can't school be more like camp?

My grandson Milo (age 9) came back from four weeks of summer camp last week. He was not happy. He loved camp and didn't want to leave. He said now all he had to look forward to is another year of boring stuff until camp starts next summer.

I asked him what he did in camp that was so much fun. He loved playing around with a whole new set of friends. And he loved the activities. He loved that he could choose to do whatever interested him. He loved the freedom. What interested him? Water skiing, archery, crafts, and the various all camp competitions in a variety of sports. He told me that when you choose an activity they set goals for you. He told me he got to level 3 in radio. I asked what that meant. He said he learned to operate the equipment, make a jingle, and put together a music show. He was very proud of himself.

I asked if they set goals for him in school. He said "school is a lot of test prep." "Camp is more free." At camp he is active. "School is more strict."

Finally he said: "We need to do something other than sitting around and test prep. Most of the year is test prep. What we learn is how to prepare for the test."

Congratulations New York City School System. You have taken a child who is bright and eager to learn and made him into someone who would do anything to avoid the boredom and absurdity of school. What a great way to educate people.

(I might point out that Milo learned about communication, writing, physics, mechanics, and a range of other "school subjects" while learning to achieve his goals in camp.)

Chapter 10: How To Fix Education

It's time for a change

The other day I had what is a rather typical conversation for me. I met someone at a bar and we chatted about various things. He told me he had just retired from the fashion industry and he told me where he went to college. He asked about me, I said a few general things but he pushed for more information. So, I said, "Actually I am trying to overthrow the education system."

This kind of remark usually gets a rather shocked response. People get defensive about their schooling. I have learned to anticipate that, so I added, "What did you learn about fashion or business in college? I am pretty sure that everything you learned in your career that was important to you, you learned on the job and that your college education had no relevance at all."

He responded, as most people do, by saying that his college education provided him the basics, which I always take as a kind of religious remark. They don't really know what "the basics" are (apart from reading, writing, and arithmetic, which he would have learned by the fourth grade) but they don't want to believe that their schooling was a waste of time. He insisted, so I asked what I usually ask in this situation: tell me the quadratic formula. He responded that he knew it but of course he didn't. Of course, he had learned it and retained it long enough to get a good score on the SAT. Then he forgot it and never thought about it again.

When we imagine school, we imagine sitting in a classroom and listening to the teacher. We recall writing papers and cramming for tests. College has the additional feature of being fun. Not because of the classes, most of whose names we cannot remember, but because of the outside activities and the people we met there.

Why do we accept a broken education system? Why do we accept that college will not teach us life skills or job skills and that school before college will be an experience that most people would never even consider repeating? Most kids would skip school if they possibly could. Why do we force them to go? And, when we force them to go why do we force them to sit still and listen? And when we force them to sit still and listen why do we force them to listen to simplistic depictions of history or read books that do not interest them or force them to do math they will never use in their real lives?

We do this for reasons that have to do mostly with day care and history. Few parents want the responsibility of taking care of their kids all day. We like government provided day care and really need it if we want to work. But why don't we ask questions about what is taught and how teaching takes place? Because the same stuff happened to us. We accept it as part of life. Maybe we even enjoyed some of it when we were kids. ("I always liked history." Really? You liked it better that your favorite childhood activity? "Well no, I meant I liked that subject best in school.")

Somehow we accept school as a painful experience of no real relevance to our lives and we talk about what we liked when we mean what we disliked least. And we talk about school's relevance by assuming it provided "the basics" when we really do not know what the basics are.

"School taught me how to think." And you didn't know how to think before? And life after school hasn't taught you how to think either? Every experience teaches you how to think better. School does it least well of all since your other experiences typically would relate to your interests, needs, goals, and achievements.

What should we learn? Certainly not the "core subjects" in school. Not only shouldn't there be a Common Core, there shouldn't be any core at all.

Why not?

When you were a child, before school, did you like to do what your sister (or brother) did? Or, did you choose to do things that interested you? Did your parents force you to study and learn certain things when you were four or five, or did they offer things, some of which you found interesting and wanted to do more of?

People are born with certain natural interests. An interest is a terrible thing to waste. The job of a parent (or of a teacher if they have that freedom) is to help a child follow his or her interests. If your daughter likes dolls you buy her a dollhouse and talk with her about what is going on in her doll house. If your son likes football you teach him to throw and catch and you take him to a game if you can. None of this is radical stuff.

Then school happens and everyone has to be doing the same stuff at the same time. And everyone has to sit still. (I never saw a six-year-old sit still unless he or she was forced to do so.) And no one can talk out of turn. These are the same kids who when out of school, run around and yell, ask questions, draw pictures, build buildings out of blocks, try to learn to swim, and so on. But school is about discipline, which is another way of saying the stuff they teach is boring and everyone has to shut up because there are too many kids in the class.

Of course, somewhere along the line, kids do have to learn discipline and how to behave in public, but this needn't be a constant all day painful lesson. And kids should be together, not in groups of 30 perhaps, but in smaller groups, so that they can learn to function together and make friends. I am not suggesting that kids stay home and learn on the computer, although that is fine with me. There should be teachers enticing them to learn something different and helping them when they have difficulties. But that does not mean that there need to be classrooms nor the inability to speak one's mind.

It is time for a change. We do not have to accept that the schools we have always had are what we must have now. Times have changed. Now everyone goes to school and now we have computers and the internet. The possibilities are endless. The economics of school can be quite different than what they are now. We can let kids learn what they want to in the way that works best for them. We will have a happier and a better functioning society because of it.

A conversation about learning

Learning is profoundly misunderstood by the school system. People learn all the time, typically when they are trying to accomplish something and are having some trouble getting what or where they want.

They may have to ask for help. That's learning.

They may have to think hard about what is going wrong. That's learning.

They may have to recall some prior similar experience and figure out its relevance to the current situation. That's learning too.

But when someone is talking to you about a subject that that person has decided you need to know about, that is not learning or anything like it.

We typically call that "teaching" but it really isn't teaching at all. It is something teachers do. It is something tour guides do. It is something drill sergeants do. It is something that leaders of organization do. But it is not teaching. It is talking at people and hoping they are listening. But, they usually aren't.

What if they were listening? Would that be learning? Probably not.

Notice I didn't say absolutely not. It is possible to hear somebody say something and learn something. Just today I asked where something was, was pointed in the right direction, and now I know where it is. Today that is.

By next year I will likely have forgotten where that place is unless I continue to go there regularly.

We only learn by listening under two conditions:

1. We continuously practice or rehearse what we have learned.
2. We didn't really need to retain what we learned so we only remembered it temporarily.

The first of these conditions I will call *real life learning.*

The second of these conditions I will call *school.*

What is the difference between real life learning and school? In school we learn things we are likely to never practice after school and thus are unlikely to retain in our memories. We might randomly retain some of it, enough to answer a question on Jeopardy or in a game of Trivial Pursuit, but we don't need it and so it is not part of real life learning.

In real-life learning we learn how to do things, usually things we will need to do again. We are not attempting to retain information, although that may happen, we are simply trying to attain new a skill like driving a car or selling or drawing up plans or designing a house or programming a computer. Schools don't usually teach useful skills until graduate school, although they may not teach them then either.

Learning happens when we try something, practice something, make something, use something, respond to something, change something, fail at something.

But how does learning take place actually? What is the medium of learning?

The answer is very simple indeed. We learn through conversation.

Why?

Because a conversation only happens when two people both want to participate. This immediately differentiates it from school in which there is only one willing participant – typically the teacher.

In order to understand what I mean here we need to grasp that conversation takes place not between two people, but between two memories. Learning happens if the memory of at least one of the participants is altered in some way. But in order for that to happen, the memory of the other participant had to be part of the process. Whatever the first memory retained had to be in the memory of the second participant in the first place.

Well, not really.

It is possible for two people to both have their memories altered by a conversation in the sense that they both come to a mutual realization about something that neither of them fully understood in that way prior to the conversation

It is also possible for the memory of one person to be changed by a conversation due to thoughts initiated by the conversation in that person's memory that were not there before the conversation started but were not in the memory of the other person at any time.

The first of these situations we call **knowledge transfer**. It is typically what we think of as teaching although neither participant in the conversation may see it that way at the time.

The second of these situations we call **mutual inquiry**. It is typical in research and intellectual conversations when both participants are trying to figure something out and attempt to do so by talking to each other about what they are thinking about.

The third if these situations we call **reflection**. Often during or after a conversation we come to realize something we had not realized before. Reflection is an internal process but it is quite often initiated by conversation.

This is how learning happens. It happens through conversation and involves the memory of individuals that are altered in some way by the conversation.

This could happen in school of course, but typically it doesn't happen in classroom.

In conversation, what we hear reminds us of something we have already experienced. From this reminding, we make responses. We may change the other person's idea. We may use their idea to tell our own story. But all of this is non-conscious. We don't know what we will be reminded of in a conversation. We don't know what we will say next. We don't control our thought process. Conversations with other people initiate our thought process by inciting reactions and ideas that we feel the need to try out on others. We need to find out what we think. We need to talk. We need to respond. We need to defend our ideas. We need to come up with ideas. In conversation, we are certain to learn.

Conversations are often contentious. That's not the only kind of conversation that works for learning, but it does work well. When people are passionate about what they are talking about, especially in the context of a project or problem they are working on, the world opens up for them.

A conversation with someone wiser than you, someone who takes time to listen to you will make you wiser. A conversation in which you must struggle with what you think, where you must defend your point of view will make you think more carefully. Conversations matter when you are discussing things that are important to you.

Despite how natural and essential conversation is to living and learning, we have neglected its power and importance in school. School has become primarily about facts and tests. School used to be a conversation. The Oxford Tutorial system was about conversation. Plato wrote about conversation. Even the Bible is all about conversation. But today school has very little real conversation.

People are having fewer valuable, challenging conversations in their daily lives. Turning off the ubiquitous noise of messages, tweets, and postings to allow the time and space for real conversation and non-conscious thought is rare. A series of 140-character remarks, no matter how clever, is not a conversation. Even a so-called discussion on social media is not likely to challenge, support, or even provoke a person to come up with new or better ideas. Nobody ever seems to post, "Wow, you're right. I never thought about it like that before."

Interestingly, when we are mentoring our most advanced students (PhD candidates), we seem to recognize the absolute and essential value of conversation in the learning process.

PhD students regularly have conversations with their advisors to discuss their thesis progress. They talk at length with their mentors about their problems and ideas in the context of their work. Some might argue that this is a valid learning method for PhDs because they've completed years of rigorous knowledge acquisition and are therefore prepared to engage in conversation involving their own questions and ideas.

So, is it only the academically accomplished who should be learning from conversations? When we are parenting, we unreservedly accept that conversation is the primary tool we use for teaching as our young children encounter the world with questions, ideas, beliefs, problems, discoveries, experiments, and fears. We recognize conversation's paramount importance in the development of children. And yet we neglect it in our schools.

What is it about conversation that matters so much in learning? Don't we learn just as much from reading a book or listening to a lecture? Well, no. You would learn more from talking to me than you will from reading any book I have written. Why? Because you would be able to argue back. I might learn something from you as well.

Consider the last time something interesting happened to you. What was the first thing you did when that experience was over?

You have a choice when something interesting happens to you. You can sit and think about it (having, in essence, a conversation with yourself). But, usually if another person is available with whom you could discuss your experience, you choose to have a conversation. But you don't tell just anyone. You find a person who will empathize with what just happened, who will help you think about it, who will challenge your assumptions, or who will just be very interested for some reason. No matter the kind of reaction you get, you benefit from that reaction as you fully digest the experience.

Conversation helps us think through what we have experienced, even if it's the experience of reading a book or watching a movie. Put another way, the only lasting benefit we receive from reading and listening or watching (aside from the entertainment value, of course) happens because of the conversations that we have about our reading or viewing experience. The learning – the changed perspective, the improved ability, the new idea – if it happens at all, happens when we are in the conversation. And as it happens, because we are humans and naturally want to learn, conversations beget more conversations. This is how human beings work. This is how we operate. And yet most school does not involve nearly enough conversation, so not nearly enough learning takes place.

Of course, some schools pay lip service to the idea of conversation. Even the MOOCs (Massive Open Online Courses) that have dominated recent news coverage about education are now starting to include so-called discussion groups. And yet with thousands of students, how can the mentors individually challenge their students and engage in meaningful conversation? Most people will admit that MOOCs are terrible, for this reason and more. But most of us have not noticed the stark absence of real conversation in the classroom. And even if we have noticed the absence, we have not recognized what it means. Without real conversation, there can be very little learning going on.

Learning depends upon conversation. Learning is fundamentally a conversation.

I'm not just lambasting MOOCs as bad. I'm saying that conversation is pretty much all that should be taking place in education and lectures and therefore MOOCs are the worst of what education has to offer. Of course, in learning conversations, you have to have something important to talk about. Ideally, you should be working on something challenging and talking about your ideas, your thoughts, and your problems.

Learning unquestionably depends upon our fathers and mothers and all the other parent surrogates and mentors who care about us enough to make the time to think and learn with us, in conversation.

Here is a conversation I once had with my father.

I showed him the latest book I'd written. He said it was unimportant. I asked how he knew that. He said that the general public wouldn't read it, so it wouldn't matter. I said that I was a professor and I wasn't writing for the general public. He said that my work wouldn't make any difference then. I was provoked, as usual, by my father. I was angry. I believed I didn't have to write for the general public to do work that mattered.

This conversation has carried on, even though my father isn't around anymore to disagree with me, because that's how people think. And I think maybe my father was right.

We learn by talking

Wait, haven't I always been the guy who said we learn by doing?

Of course. Talking is a kind of doing. But that is hardly my point. Plenty of academic courses insist that they use learning by doing as a methodology. After all, writing an academic paper is a kind of doing too. So, one would learn how to write an academic paper by writing one. But, some clarification is needed as to what kind of learning matters. Is it important to learn to write an academic paper? It is if you plan on becoming an academic.

But the real question about what high school or college should be like should be centered on what learning is like, apart from what is actually being learned. We need to understand what the fundamentals of learning really are.

Many years ago I was having lunch with my closest colleague. I was complaining to him about the way my wife cooked meat. It was always too well cooked for me. He responded by saying that fifteen years earlier he had tried to get his hair cut in England and they wouldn't cut it as short as he wanted it.

It seemed like an odd response, so I spent some time thinking about it. What did haircuts have to do with rare meat?

At one level nothing. But, at a higher level of abstraction these were identical stories. We both had asked someone to do something for us that they were capable of doing, but they had refused because they thought the request was too extreme.

Instead of focusing on why my friend was peculiar because he had answered the way he did, I assumed something interesting was going on in his head and attempted to figure out what had happened. I wound up focusing on his need to reconcile a failure he had had years ago. (We remember our failures.) I focused on how the process worked. In a short time, I had come up with a theory of how memory is organized (around stories indexed by abstractions such as "refusal to satisfy someone else's goal"). This began a long process that I still work on, together with many students and colleagues, to get computers to self-organize their memories. (We work on this by talking about it.)

Thinking by oneself is hard because there are too many distractions. I noticed that I would wake up in the middle of the night with ideas and I wondered how that was happening. I realized that our non-conscious mind does all the thinking (my friend wasn't consciously looking for his haircut story after all, it just showed up in his head). I began to realize I could let my non-conscious self do my thinking and then later consciously recognize what I had thought.

Conversation is a non-conscious act. We don't know what we will say next. We don't know what we have just heard will remind us of. And we don't control our thought process. Conversations with other people enable our thought process to begin by inciting reactions and ideas that we feel the need to try out on others. We need to find out what we think.

Turning off the noise to allow non-conscious thought was hard when I was working on this problem 35 years ago; it is twice as hard today. Your phone is always available, your computer is nearby, and there might be a text message, or a tweet, or a Facebook posting. But those are not conversations, although they may look that way to people at first glance.

A series of cute remarks is not a conversation. Even a discussion on Facebook, while appearing to be a conversation, is not likely to challenge one to come up with better ideas and quickly think new thoughts. I have yet to see a Facebook comment that said "You are right, I never thought about it like that before."

What does this have to do with learning? Dialogues were used by Plato to discuss how learning works. But one needn't go back that far in order to understand that the true relationship between teacher and student ought to be one of dialogue.

We see this easily when we consider graduate education. Students meet with their PhD thesis advisors to discuss their progress, their ideas, their problems, and then, presumably, they are ready to go back to working on what they were doing with new insights.

This same sort of thing happens between parents and small children when they ask numerous "why" questions on being confronted with new things, new people, or new ideas.

Why does conversation matter so much for learning? Couldn't you just read a book or listen to a lecture? Wouldn't you learn from those experiences as well?

Well, no.

To explain what I mean here, consider the last time something interesting happened to you. What was the first thing you did when that experience was over?

People really have only two choices when something interesting happens. The first is to sit and think about it some. To have a conversation with oneself in other words. But it is rare to choose that option when there is another person available to whom you could tell about your experience. That person has to have some qualifications of course. You can't just tell anyone. We find people to talk to who will empathize with what just happened, or who will help us think about it better, or who will challenge our assumptions, or will just think what you have to say is wonderful. No matter the kind of reaction we get, we need that reaction. We must tell our story, even if it is just a story about a movie we just saw, a book we just read, or a lecture we just heard.

To put this another way, the only reason that reading and listening aren't totally useless experiences from a learning point of view (they might be good entertainment of course) is the conversations that they spark later. The real learning takes place in the conversation that follows. And, conversations follow conversations. You hear someone say something and you repeat it to someone else and discuss some more. This is how human beings work. But it is not how school works.

Of course, schools pay lip service to the idea of discussion. The MOOCs that have dominated recent conversations about education have discussion groups for exactly this reason. But, these discussions are not one-on-one with the teacher. (How could they be with thousands of students?) So MOOCs are taking the very thing that is most needed in education, one-on-one conversations with the teacher, and eliminating their possibility.

My view is more radical than just that MOOCs are bad however. If learning is fundamentally a conversation, then conversation is all that should be taking place in education. Well, not all. You have to have something to be talking about. You should be doing something and talking about what you are working on. Learning is a conversation. We need to get rid of classes (unless they have fewer than ten people and are really a conversation), tests (which are the antithesis of conversation), and any other aspect of school that does not involve learning to express your ideas and have them dissected and responded to by interested parties who can help you make your own ideas better.

Persuasive arguments

Teaching students how to move people to their point of view is a very important thing to do. Challenging students to try to persuade fellow students by debating in public for example, is a very useful thing to do in education. It is useful because constructing and backing up arguments causes you to think hard. The more you have to think hard the better you get at it.

So, I have a simple suggestion for school. Teachers should stop having persuasion conversations all together (where they are the persuader) and help students learn to persuade each other better. Students learning to persuade is a very valuable educational goal. We need to make that part of any school we create.

But, of course this is very difficult to do within the current system. Here is an article from the *New York Times*:[1]

> Texas' State Board of Education has approved new history textbooks, but only after defeating six and seeing a top publisher withdraw a seventh — capping months of outcry over lessons that some academics say exaggerate the influence of Moses in American democracy and negatively portray Muslims.
>
> The board on Friday approved 89 books and classroom software packages that more than five million public school students will begin using next fall. But it took hours of sometimes testy discussion and left publishers scrambling to make hundreds of last-minute edits, some to no avail. A proposal to delay the vote to allow the board and general public to better check those changes was defeated. "I'm comfortable enough that these books have been reviewed by many, many people," said Thomas Ratliff, a Republican and the board's vice chairman. "They are not perfect. They never will be."
>
> The history, social studies and government textbooks were submitted for approval this summer, and academics and activists on the right and left criticized many of them. Some worried that the textbooks were too sympathetic to Islam or played down the achievements of President Ronald Reagan. Others said they overstated the importance of Moses to America's founding fathers or trumpeted the free-market system too much.

Why does the government think that it should direct the conversation? The government is after all just an assortment of politicians with what is probably a rather limited view of history. The answer is simple enough. Politicians understand persuasive conversation well enough, and they want to direct it. They could, of course, simply participate in it, allowing others with different points of view to participate as well. But they don't. Politicians see school as way of indoctrinating students, and they always have. If we are

1 Texas Approves Disputed History Texts for Schools http://www.nytimes.com/2014/11/23/us/texas-approves-disputed-history-texts-for-schools.html

ever to change our schools to ones that teach thinking, we must allow students choice in what they learn, and choice in what they choose to believe. We must encourage them to reason from evidence and not from someone older who wants to tell them what to think. This is not easy to implement.

The kind of thing we see happening in Texas here, happens in one way or another everywhere. "Truth" ought not be taught in schools. Students need to learn to verify, not memorize.

What should a persuasion conversation be about? How should one be conducted? How can we help students be persuasive?

Instead of teaching history, how about if we asked students to convince other students why it was important to learn history and what history it was important to learn? Instead of politicians having that debate (not really, they all know the answer) let's let students have the debate.

This week's assignment: *Was Moses important to America's Founding Fathers?* How could we find this out? What evidence is there? Why would it matter if it were true? Who benefits from believing it was true? What would happen if it weren't true?

Next week's assignment: *How good a President was Ronald Reagan?* How can we know if a President succeeded? What should the criteria be for success for a President? Whose interests does it serve to have Ronald Reagan be seen as a great President?

Another assignment: *What is the free market system?* Who wins? Who loses? Why does the Texas School Board care about this?

Now I am making a simple point here about persuasive conversation. It can be about anything. But students need to be involved in making judgments of the sort the Texas Board is making. They should be in this conversation, not for political reasons but because it is within such conversations that real thinking takes place. While no real thinking probably goes on in any actual Texas Board meeting, students would not be serving vested interests when they addressed those issues and would not be making any real decisions anyway. They would just be learning how to be persuasive using evidence, facts, and reasonable argumentation. They would be learning how to attack and defend such arguments in a reasonable way. This is what learning in school or out of school looks like, or should look like.

Students need to be in persuasive conversations in order to learn.

Change just one thing

Last week I was interviewed by phone from Spain. I was talking to authorities who were preparing a report for the King of Spain on how education might be improved in Spain. I am well-known in Spain so it is not odd that they were calling me. They were certainly calling many others as well.

I started by saying that I am really radical and they said they already knew that. I then talked with them for about a half an hour about the kinds of improvements to education that I have been writing about for years in my columns and of course in my latest book, *Teaching Minds: How Cognitive Science Can Save Our Schools.*

They seemed to be enjoying talking to me and hearing what I had to say. Then, they asked one final question: "If you could just say one thing that needs to be changed, what would it be?"

It is easy to imagine that they wanted a one-liner for an executive summary. I don't think I gave them what they wanted, judging from their reaction.

I said, "Just eliminate classrooms."

They audibly gasped.

Why?

First why did I say it?

Because if you eliminate classrooms everything else follows. No teacher talking to kids who aren't listening. No tests to see if they were listening. No kids distracting other kids who are bored by what is going on. No subjects that don't relate to the interests of the child. Instead, without a classroom you can re-invent. We can think about how individuals can learn and while doing that we would need to confront the fact that not all individuals want to learn the same things. We would have to eliminate the "one size fits all" curriculum. We would need to create curricula that met kids interests. We would be able to let kids learn by doing instead of vainly attempting to have them learn by listening. We could eliminate academic subjects. We could make learning fun. Classrooms are never fun.

Why did they gasp?

Because they can't do it. They knew it and I knew it. They don't really want to fix education. They want to make schools function better. And schools have classrooms. And that, my friends, is the beginning and end of the problem.

The old university system is dead - time for a professional university

I once had lunch with a member of the Board of Trustees of the University of Illinois. I asked him how it felt to be in charge of a fraudulent institution. He was shocked by the question, of course, but I continued. How many of the people who attend the University of Illinois do you think go there because they think they will get jobs upon graduation? He supposed that all of them did. I asked if they actually did job training at Illinois. He agreed that they didn't. I pointed out that most of the faculty there had never had jobs (except as professors) and might not know how to do any other jobs. He agreed.

I once suggested fixing this state of affairs while I was a professor at Yale. I discussed this with the President of Yale at the time, Bart Giammati. He replied that Yale didn't do training. But Yale does do training. Yale trains professors.

Universities were started as places for classical education for wealthy people. Typically they indeed had a large training component since they were usually religiously based and future religious leaders were expected to come out of these schools. Yale started as a divinity school and morphed into a school where the wealthy could hobnob with members of their own class of people and then turn out to be the leaders of the country (who knew Latin and the classics). Although most did go into business, business was never taught.

The world has changed in the sense that anyone with exceptionally good test scores and grades can get into Yale. But Yale hasn't changed. One is still expected to study the classics and there are no programs for job training. Even the computer science department, of which I was part for 15 years, has no interest in training future programmers.

I am not criticizing Yale here actually. Most universities have copied the "training of intellectuals and professors model of education" and have disregarded the idea that future employment might be of major concern to students. Professors can do this because they are forced by no one to teach job skills. They don't really know much about job skills in any case. The major focus of a professor at any research university is research. Teaching is low on their priority list and teaching job skills is very far from any real concern. So, economics departments teach theories of economics and not how to run a business, and law schools teach the theory of law and not how to be a lawyer, and medical schools teach the science of the human body but not how to be a doctor. Psychology focuses on how to run an experiment, when students really want to know why they are screwed up or why they can't get along. Mathematics departments teach stuff that no one will ever use, and education departments forget to teach people how to teach.

Still we hear that everyone must go to college. Why?

It is time for a change.

I propose the creation of a Professional University. By this I mean a university that teaches only job skills. It would do this by creating simulations of the actual life of someone who works in a particular job. After a year or more living in a simulation of the life of an actual engineer, computer scientist, psychologist, business person, or health care professional, the graduate would actually be employable and would have a pretty good idea of whether they had made a good choice of profession. Creating these simulations is not that complicated. Small projects can lead to larger projects that build upon what was learned in the earlier projects. Constant required deliverables with mentoring by faculty, not lectures. Students try to do things, with faculty there to help.

The faculty in a professional university would be practitioners who had done it themselves. The students could come to campus or work online. It makes no difference. Deliverables would not be given grades. If your business idea isn't good, work harder on it. If your technological solution to something doesn't work, keep working on it. Degrees would not be based upon an accumulation of credits and would have nothing to do with the time a student was in the program. The students would have to complete well-specified tasks, and when they demonstrated they could actually do something they would move on to a more complicated task. If the students weren't immediately employable, the programs would have to be modified until they were.

We can do this. It just takes money. Existing universities won't help. They will be threatened by it. Yale can keep producing professors and intellectuals. But most countries in the world need way fewer professors than they need well-educated functioning professionals.

Would graduates of the Professional University be able to speak, write, reason, and solve complex problems? Of course. Those skills would be built into every program.

What cognitive science tells us about how to organize school

We have all gone to school. We all know that school is organized around academic subjects like math, English, history and science. Why?

It is not easy to question something that everyone takes for granted. It is especially not easy when the very source of all our concerns in education can be easily traced to this one decision: to organize school around academic subjects. How else might school be organized? There is an easy answer to this: organize school around thought processes. In 1892, when the American high school was designed, we didn't know much about thought processes. Now we do. It is time to re-think school.

School, at every age, needs to be designed around these processes, since it is through these processes that everyone learns. Academic subjects are irrelevant to real learning. They are not irrelevant to the education of academics of course. But, how many people really want or need to become experts in academic fields?

Here is a list of the sixteen critical thinking processes. These processes are as old as the human race itself. The better one is at doing them, the better one survives.

The Sixteen Cognitive Processes that Underlie All Learning

Conscious Processes

1. **Prediction**: determining what will happen next
2. **Judgment**: deciding between choices
3. **Modeling**: figuring out how things work
4. **Experimentation**: coming to conclusions after trying things out
5. **Describing**: communicating one's thoughts and what has just happened to others
6. **Managing**: organizing people to work together towards a goal

Subconscious processes

7. **Step-by-Step**: knowing how to perform a complex action
8. **Artistry**: knowing what you like
9. **Values**: deciding between things you care about

Analytic Processes

10. **Diagnosis**: determining what happened from the evidence

11. **Planning**: determining a course of action
12. **Causation**: understanding why something happened

Mixed processes

13. **Influence**: figuring out how to get someone else to do something that you want them to do
14. **Teamwork**: getting along with others when working towards a common goal
15. **Negotiation**: trading with others and completing successful deals
16. **Goal Conflict**: managing conflict in such a way as to come out with what you want

All of these processes are part of a small child's life as well as a high-functioning adult's life. Education should mean helping people get more sophisticated about doing these things through the acquisition of a case base of experience. Teaching should mean helping people think about their experiences and how to handle these processes better. Unfortunately education and teaching rarely means either of these things in today's world.

Start by asking students what they want to learn

I am in London as I write this. I have been riding the trains to get to places like Brighton and Sunbury for business meetings. I love riding trains.

Now, ordinarily the fact that I love trains would be of little interest to anyone, but there is more to the story.

Some years ago, when I was trying to get my father, who was over 80 and visiting me at the time, to do something he didn't want to do, I told him we could ride the Chicago subway to get there and he immediately agreed.

OK. So my father and I both like trains. I loved riding down to Florida when I was a kid and waking up in Jacksonville after an all-night trip from New York and seeing the sun shine and feeling warmth everywhere. My father and I rode together while my mother slept in a sleeping compartment. My love of trains started early. So just childhood unconscious emotional stuff, right?

Except both of my grandsons, ages five and three as I write this, love trains. Actually obsessed with trains is more like it. One lives in New York City and the other in Washington D.C. They each know every train and route in their respective cities and generally demand to watch trains when I play with them on Grandparent Games.[2]

Is there a train-loving gene? Certainly it would have to be a very recent mutation, so it is a silly idea. And besides, my daughter, whose son is the five-year-old in New York, never seemed to be fascinated by trains.

Of course, I left out my son, the one who has a PhD in transportation and runs a transportation policy think tank in Washington. My son was so obsessed with trains as a kid that when I showed him the Paris Metro when he was 10 (we had just moved there for a year) he said, "Why have you been keeping this from me?"

Train gene or not, the point of this story is to talk about education of course, and to talk about how school needs to be re-structured. My son did fine in high school but he wasn't passionate about much. He decided he wanted to be a history major when he arrived at Columbia University as a freshman. (He chose Columbia because there were trains to ride there of course. He almost died when I suggested Cornell or Princeton.)

I was (and am) a non-typical father, one who always felt happy to direct my children's pursuits and one who was a college professor and knew a bit about universities. So I told

2 http://www.grandparentgames.com

him history was off the table as I saw no point in studying it, and that he should major in subways. He was shocked. "How do you major in subways?" he asked. I said I was sure there were people who did transportation at Columbia and to find them. He signed up for a graduate seminar in his first semester there (putting off a required humanities course) and figured it out from there, later going to MIT for a Masters in Transportation and returning to Columbia for the PhD.

My son loves his work because he is, and always was passionate about trains (and later on, planes).

Schools need to allow children of any age to follow their passions. Educators need to stop telling students what they should learn and should start asking them what they want to learn. How crazy an idea is that?

As for the genetics I don't care really. But there is solid male line of train-loving in my family.

My son has recently taken the job of Chief Innovation Officer of the Los Angeles County Transportation Department – good luck with that!

Pro-Choice: here's how to fix high schools

I have been writing about high school and what is wrong with it for many years. My articles on why all the subjects we teach are absurd and why the curriculum is tremendously outdated are easy enough to find. Often people respond to what I have written by asking what we should have instead. So, here I propose a simple answer. One we can implement, and one we can gradually get into the schools.

Pardon me for calling this Pro-Choice (by which I mean professional choices for kids). Yes, I know the term means something else. But I like it in this context.

My premise is that high school should be a time in which one figures out what kinds of things one can do in life that would be just right for you. This idea has been around for a long time but used in exactly the wrong way. "We must teach chemistry in high school so we can expose children to chemistry to see if they want to be chemists" is the standard argument. It didn't take me a year of high school chemistry to know I didn't want to be a chemist. Had we had what I am proposing, my decision would have been even easier. (And I had to take two years of college chemistry too. Believe me, I knew long before then, but schools just love requirements.) In the school I am proposing there are no requirements – just professional choices.

I happen to have spent some time with a chemist at Proctor and Gamble a few years back. He was inventing a new bleach. Let us imagine for a moment that Proctor and Gamble funded the building of a three-week chemistry learn-by-doing experience that included seeing what chemists actually do at P&G, talking to this man about why he loves what he does, and actually doing some of these things in simulation. After three weeks a student would know if this was for him or her and if they wanted more of it, or if they wanted to try out something else.

Years ago we built a simulated firefighter course (at Northwestern's ILS). Suppose we allowed high schools kids to try out being firefighters in simulation for a few weeks. They might even talk to their local firefighters during that same time. In those days, we also built simulations about how to run an EPA public meeting and about how to plan an Air Force campaign. If we build a version that kids could try, they would know if that kind of career was for them after a short while.

You say you want to be a lawyer? Why not try a case in simulation? Do contract work too, to see that being a lawyer is not all *Law and Order.*

You are thinking about being a doctor? Be one in simulation. Talk to simulated patients. Do some lab work. Read an MRI. Tell a patient he has cancer (all in simulation of course). Kids could help out in a real local hospital for a few days.

Why shouldn't GE help us build a three-week simulation of what it is like to be an engineer? Why shouldn't IBM help us build a simulation of what it is like to be a computer consultant? Why shouldn't one of the political parties help us build a simulation of what it is like to be legislator or a campaign director? Why shouldn't Turner Construction help us build a simulation of what kinds of jobs there are in construction and see if they'd be any fun to do?

I am naming particular companies here because I believe the only way education will change is if big corporations – which can easily afford to help us do this and would benefit from it – helped provide students with choices.

How many should there be? Hundreds. A student's life could simply be trying stuff out, talking to experts, and going on to the next until they were pretty sure about what they wanted to learn more about.

We have built many of these already. Many of them are in health sciences and in computer programming and in entrepreneurship.

Now. How do we get them to the kids?

No one will allow us to eliminate the nonsense that permeates high school today, but there are electives available to seniors and there are summer schools, camps, and after-school programs. Eventually maybe we can eliminate the entire last semester of high school and replace it with simulated activates that inform kids about what they might like to try in the future.

My long-term plan, of course, is have this *become* high school, gradually replacing what is there.

What is there can easily go. If you actually needed algebra you could learn it in context. It might be embedded in an advanced engineering simulation when a student was building a bridge or designing an airplane. (Although I must admit I had this conversation with Boeing for high school aerospace engineering and they couldn't find a real need for algebra there either.)

English literature? Needed for nothing (except sounding like an intellectual). However, as literature teachers would say that literature is about making life decisions, I have no problem with many "making life decisions" simulations being part of the choices here. Writing is needed all the time however, so each simulation should involve writing that is the type actually done in that simulation such as legal briefs, medical opinions, police reports, etc.

All of education needs to involve planning, diagnosis, judgment, predication, and experimentation (as I have said in my *Teaching Minds* book). These cognitive processes must be woven into each and every simulated experience we build.

High school must change. Computers and the internet allow us to make the change now. We need to think about enabling choices for students and creating individuals who know what they want to do and have tried it out before they finish high school.

We have built many of these kinds of simulations already and will offer them to anyone who wants to try them. We need money to build more. (All of this is being done at my non-profit, Engines for Education.)[3]

3 http://www.engines4ed.org

High school: Out with the old and in with the new

What might high school look like if we really thought about re-designing it in a serious way? By this I mean, in a way that ignores what textbook makers, test makers, Common Core advocates, and teachers who do not want to change how they teach want.

Or, to put this another way, how can we make high school, fun, exciting, useful, and something that sends children off on a path that reflects their own interests and passions?

We need some clearly defined outcomes first, so let's state upfront that there are some core skills that must be learned in any curriculum but that these are not the ones that we usually talk about when we go through the usual litany of mathematics, science, history and literature.

I assume, therefore that for any curriculum I discuss below, there will be a heavy component of reading, writing, teamwork, and reasoning. And, I assume that reasoning would include planning, prediction, judgment, evaluation, and other core cognitive skills I have discussed in the past. (*Teaching Minds*, Teachers College Press)

The First Year of High School

The goal is to get students excited about something. This means that students would be offered the option of working on projects with clearly defined goals in the following areas:

Science, engineering, design, art, music, health, construction, architecture, computers, business, law, finance, anthropology, philosophy, history, psychology, film, television, foreign languages, foreign cultures, service industry.

This is not meant to be an exhaustive list. It is meant to reflect the range of jobs one can have in the world. We can always make it bigger. Some of the terms above are quite general. So, by "business" one could mean wholesale, retail, investment banking, insurance, and a range of other things. The idea is to enable a student to do any of those that a student chooses to do.

I am proposing that the list be finite. So, for sake of argument, since I listed about 25 domains of interest, let's say that a student entering high school would have to make a choice to pursue one of these 25 areas for one month. That month would consist of one project, with the material for it online, with an online mentor available, and with a physically available teacher watching to see that students were engaged and working and available to help when they were frustrated. The students would work in online teams of 3-5 kids, who could be located anywhere. The projects would not be teaching theory, just practice at doing something simple within that domain. They would all involve writing, drawing conclusions about how to do things, reflection, and discussion. There would be no grades. At the end of the month, the student would have a simple choice to make:

1. Leave this school and do something else.
2. Do a next project in the same domain that builds on the one that was done in the first month.
3. Change domains and do a different project.

The first year of high school therefore would have no classes, no tests, and no grades. It would have lots of choices. Eight months of high school could mean eight unrelated projects, or one project area that gets increasingly complex each month, or anything in between.

The Second Year of High School

The student would be encouraged to change the game plan that he or she has followed so far. So, for example, if the student did only music, or only computers for the first year, they would be encouraged to choose something else to concentrate on, but would also be allowed to pursue what they had started in parallel. The point here is to make sure that a student doesn't get too narrow too fast, and to allow students who are excited by something to continue to pursue it.

The Third Year of High School

By now, a student would have tasted seriously at least two or three domains. At this point they could choose to pursue two of them seriously, or they could continue trying our new things.

The Fourth Year of High School

In the student's final year they choose one thing and stick to it. They must produce something worthwhile or invent something or demonstrate the ability to be useful to an employer in some domain. Businesses would be encouraged to hire students as interns to try out the skills that would by this time have been honed for 1-3 years in a given domain of interest.

What would this high school produce? Happy, employable kids who could choose further study or simply go to work.

How hard is this to do? It simply requires money to build it and help from experts in doing the building.

We can do this. We simply need to abandon the old model and get started.

Cash in those chips

It is always nice to visit with people who have a point of view different than one's own.

I have been promoting my new virtual high school, called VISTA (Virtual International Science Technology Academy). I am talking to folks who want to sponsor it, host it, or run one or more of the curricula in their own schools. The proposed curriculum, which we have begun to design, consists of four, full-year experiences in this order:

- Scientific Reasoning
- Health Sciences
- New Technology
- Engineering (aerospace focus)

These are story-centered curricula. Students work in teams in virtual apprenticeships with experts producing deliverables that get increasingly complex throughout the year. No classes. No tests. One curriculum per year – complete four of them and you graduate. Ideally there would be hundreds of curricula to choose from but we have to start somewhere so I chose those four.

When I talk to people who might be interested in radical education reform I always ask what curricula their communities might need so we can think about how to produce those as well. The idea that every high school should be more or less the same offering of the same potpourri of algebra, American history, and Charles Dickens is just absurd, so I ask what they need in their world.

The other day I visited a school for Native Americans in the Southwest. After some discussion of their needs, we came up with the following alternative to my alternative:

- Year 1: scientific reasoning
- Year 2: alternative energy issues
- Year 3: land management and forestry
- Year 4: casino management, or entrepreneurship, or tribal governance

The current high school curricula were conceived in 1892. Time for new ideas.

It's an old story - learning hasn't changed

A few years ago I was asked for my annual prediction by an e-learning magazine and I predicted the death of m-learning (mobile-learning). I was attacked by everyone. Funny we don't hear so much about m-learning any more.

Learning is a very trendy field. There is always the latest and greatest thing that everyone must do. Today this is "social learning" and "on-the-job learning."

There is one problem with this. None of this stuff is ever new in any way. Learning hasn't changed in a million years. Did I say a million? Too conservative. How do chimp babies learn? Socially? Of course. They copy what their mothers do and what their playmates do. (Amazingly they do this without Facebook.)

Do they learn on-the-job? Apart from the fact that chimps don't actually have jobs, that is the only way they learn. In the process of doing something they either fail and try again or someone helps them out.

Mentoring. Another learning innovation, except that there has always been mentoring, Parents, big brothers, helpful neighbors – all there to help you when you are in trouble. None of this is new.

But suddenly big companies have discovered it. Good for them. Better than classrooms and books (which are very new, if you think about it, cavemen didn't have either).

I play softball regularly. When I first started playing in this league I noticed a guy who was the best hitter I ever saw. I asked him questions. He gave me tips. I asked for criticism. He gave it to me. The other day I was hitting really well. I was congratulated by my team. I told them I owed it all to him. They didn't know what I meant. I said I had appointed him my personal coach ten years ago.

What confuses me is why this has to be institutionalized in big companies. It is not that complicated. Tell everyone they need to spend an hour a week mentoring and an hour a week being mentored. Let them say officially whom they have chosen. Create a culture where mentoring is the norm. It is the norm in sports. My mentor has never has asked for anything back. I am sure people mentored him over the years.

On-the-job learning is more complicated. Why? Because the right tools might not be available to do it. What are the right tools?

1. Someone to ask who can give just-in-time help.

2. A short course that one can take just-in-time and that one is allowed to take when it is needed.
3. A group that is available for discussion.

I will explain each.

Just-in-time help has always been available to most of us. It is called mom or dad. Even today I get "help" calls from my grown children. They know I will stop my day and help them. I always have.

How do we institutionalize this in the modern world? By recording all the help type stories that an expert has and making them available to anyone just in time. It sounds complicated and it is. We have built such a system. It is called EXTRA (EXperts Telling Relevant Advice). Every organization needs one. Experts move on and their expertise goes with them. Capture it and learn how to deliver it just in time in short bits that last less than 2 minutes.

Stories from experts matter. Not in the form of long lectures but in the form of a conversation that happens when there is an interest in hearing the story.

To put this another way, mentoring is not driven by the mentor. As a professor of PhD students for 35 years I served the role of mentor to a lot of people. They showed up in my office once a week because I told them they had to. After that I told them nothing. Instead I listened. Maybe I asked a few questions to get them to talk if they were shy. But learning happens when someone wants to learn not when someone wants to teach.

I did the same when I taught classes. I set up questions and listened. I encouraged students to argue with each other. I chimed in at the end when they were ready to listen.

Apprenticeship is the other side of mentoring. An apprentice takes on jobs assigned to him. A good mentor lets the apprentice drive every now and then. Surgeons let interns make the first cut after they have watched the process many times.

In the end there is always a story. In the modern era we can deliver stories when someone needs one. (When they ask or search, or we simply know what they are doing and what would help them do it.) But, the old method still works. Talking.

The problem with big companies is that they set up training sessions that last for a week instead of mentoring sessions that last for an hour. Once a week everyone should meet with their mentor for an hour and talk. Just talk. Maybe a beer would help.

And what do they talk about? A good mentor knows that the mentee drives the conversation. Maybe the mentor saw the mentee make a mistake and could comment on it, but younger people know when they are struggling and are always ready to learn if they respect the person who is helping them.

Formal training really has never been a good idea. The army does it for new recruits but they do it because they are trying to create soldiers who don't think and just follow orders. At the higher level of army training, at the Army War College for example, officers sit around and talk.

There do not have to be mentors in such situations. People who work together should have the opportunity to exchange "war stories." This should just happen late at night in bars. It is the most important training there is. But there has to be time made for it. And no it doesn't require Twitter or Facebook. Social learning has always been how we learn. It is in fashion again and that is nice but it is nothing new. The elders have always gathered around the campfire to discuss the day's events.

Do we need to teach people how to mentor and how to discuss? Yes and no. Excessive talking, lecturing and such, has never been a good idea and is never tolerated in societies that are truly cooperative. The key is learning to listen.

Listening, oddly enough, does need to be taught. Most people don't really know how to do it. They learn the hard way that listening works as they get older. Should we teach it? Yes. How?

We need to put people in situations where listening is demanded of them and where they are likely to fail to do it. (Training is one such place where people tune out. That is why there are tests, but tests usually don't test anything important.)

Having to perform is the best test.

Summarizing: Short courses delivered just in time are better than training sessions. Gathering a company's expertise and delivering it via tools like EXTRA matters a great deal.

But most of all, learning to listen and advise well is what separates winning teams from losing ones. To support these activities, an organization must formally make time for them, otherwise they won't happen. Do it on Twitter if you like.

How to fix the "STEM crisis"

This just in from the BBC:[4]

> Several universities have warned they may be forced to close science and engineering courses if the government limits visas for foreign students. Sixteen university vice-chancellors have written a joint letter to The Observer[5] saying the plans would have a profound effect on university income.

I really like the honesty expressed here. The reason universities want foreign students is so they can make money running courses that those students want to attend. The interesting part here is that the issue is science and engineering courses.

Lately I have noted President Obama's obsession about teaching science and math. Although this story is from the U.K., the lesson is the same. Either American and British students simply don't like science and engineering, or their universities have produced far too many science and engineering degree programs.

It doesn't really matter which of these is the case, as it is clear either way that the reason President Obama is saying 'science and math' nonstop is that he is getting pressure from many quarters, especially universities.

As a long-time professor of computer science, I am well aware that the vast majority of students in U.S. masters programs in computer science are from India and China. This is also true of engineering. If the supply of Indian and Chinese students were limited in the US, most university graduate programs would shut down.

I have no stake in this whatsoever, but I do have a point of view that the British and American authorities might want to listen to. The math and science programs in high school (and college too) are so awful that they deter most prospective students. The Indians and Chinese persevere in their home versions of those programs because they know that that is their ticket out. The U.S. and U.K. students have no such motivation.

We might consider building curricula that cause children to get excited about science and engineering, if that is indeed so important, by making some compelling programs. I am building a first grade engineering curriculum at the moment, not because I care about what happens in graduate school but simply because I know little kids like to build things and I think it would be fun for them.

In order to change who applies to graduate school, you will need to change high school. But high school has been the same since the nineteenth century.

4 University courses 'may close' if student visas cut http://www.bbc.co.uk/news/mobile/education-12658309

5 Cuts in the number of international students spell disaster http://www.theguardian.com/theobserver/2011/mar/05/letters-international-student-cuts

Get rid of the nonsense that is high school math and science and teach kids how to reason scientifically and how to build things, and we will see a change.

Why isn't this avenue the one that is being taken? Because it would take longer to do that than any politician's term will take. No politician ever proposes a long-term strategy. High test scores and more testing is a short-term strategy that will never achieve a good result.

Make it interesting and they will come.

To fix a country, fix education

A friend of mine went to visit Fidel Castro a few years back. (He is not your typical guy and I have no idea how this was arranged.) They got into a conversation about education. My friend mentioned me, and Castro asked whether I might want to be the Minister of Education of Cuba. When my friend told me about this, he asked what I would do if I had that job. I replied that I would ask Castro what Cuba wanted to be.

My friend found that an odd response. Some days later, Castro shot some people and the U.S. prevented my friend from visiting Castro again so that was the end of that.

I was reminded of this incident because, as I write this, I am on a Greek island and, not surprisingly, talk centers on what to do about the economy. Having recently been in Italy and Spain as well, it is obvious to me that the problems these countries are having stem from issues in education.

When I say that, the response is usually less than enthusiastic, because it seems an odd idea, so let me explain.

When I mentioned what I would want to ask Castro, this is what I had in mind. Education is meant to achieve something, although this is usually forgotten in education reform conversations. The people who designed the U.S. education system around 1900 knew this well. The country needed factory workers, so keeping students "in dark, airless places" doing mindless repetitive work, seemed like a good strategy.

Today we have the factory worker strategy still in place, reinforced by a push for standards and multiple choice tests everywhere. The fact that there are no more factories seems to have skipped people's attention. Also we have a big push for making sure everyone goes to college, despite the fact that college produces students who study what the professors happen to teach which means English, history, mathematics, philosophy, sociology, and any number of subjects that will not make students in any way employable.

In the U.S. we have gotten away with this attitude for many years because we simultaneously had a big push by the Defense Department for new technology and thus were able to create Silicon Valley and enable an atmosphere of technological innovation. So while we have no factories, we do lead the world in software. It is almost as if someone in the Defense Department in the 60s and 70s were planning this. (I was there. They were.)

Now think about Spain. Its number one industry is tourism. You would think therefore, that in Spain the schools would be pushing hospitality or cooking or hotel design. But they are not. They have their enormous share of useless language and history majors as well and the university establishment works hard to keep things as they have always been.

Or think about Greece. Their number one industries are tourism and shipping. I have been an advisor to a Greek shipowner for over a decade now, and I can tell you it isn't all that easy to learn about shipping in a Greek university. Nor is it easy to learn about tourism, because Greek universities, like those everywhere, are run by people who are worried about insisting that things stay the same so that their professorships are still relevant.

What Greece and Spain need to do, what Cuba needed to do, what any country that is not big enough to do everything needs to do, is pick its spots.

Universities offering a classical education are fine when only the wealthy elite are being educated. But mass education requires that schools be run by people who are trying to educate for the future. This does not mean educating for "21st century skills" whatever that might mean. What is does mean is that schools need to do two things.

First, they need to teach general thinking skills, not math, but planning, not literature, but judgment, not science, but diagnosis.

Second, countries need to decide what they want to be when they grow up. Cuba, had I been running the educational show there, would have had to decide what they wanted to be the best at, like biotech or agriculture or the technology of cigar making. And they would have had to offer something less than everything under the sun to their students.

To fix an economy in the long run requires planning. The planning has to start at the beginning by creating citizens who can both think and find useful employment in the sectors of the economy that the country already has or wants to have.

Education is where everything starts. Countries can simply decide to be good at something and make themselves good at it. The U.S. decided exactly that about computer science 40 years ago. But it doesn't require the wealth of the U.S. to do that. Modern educational techniques, especially high quality experiential online education, can make any country a specialist in any industry that it can realistically dream about.

A summer to-do list for students who hate (or love) high school

Every day, high school students get to my blog by typing "high school is useless" or something similar. I worry about those kids, but I also worry about the kids who think high school is very important, who study all the time, and who obsess about getting into a "good college."

The good news is that it is summertime. Now you can forget about school and actually learn something. The teacher is you. (Your best teacher will really always be you. If others in your life can help, great, but it will still be you in the end.)

So what should you do this summer? Here are ten suggestions:

1. Start a business
This is, of course, easier said than done. That having been said, my five-year-old grandson, Max, ran a lemonade stand the other day, made some money and was very excited about it. Think about how people make money where you live. Think about what services are lacking where you live. Think about what people need that you could provide. Maybe your business could be on the web and sell to people like you. What would you buy from someone like you? They don't teach how to start a business in high school. They should but they don't.

2. Learn a real skill
What constitutes a real skill? Computer programming is a real skill. Glass-blowing is a real skill. Carpentry is a real skill. Playing music is a real skill. Building something is a real skill. Pick something that sounds appealing and find out how to learn to do it. Then practice a lot.

3. Play sports
Why is this important? Because sports teaches you a few very important things that school misses. One is how to lose. Another is humility. No matter what sport you pick there will always be someone better than you at it. Sports teaches you to try harder. Sports teaches you how your body works and how to make it work better. All stuff school ignores that is very important.

4. Invent something

What is missing in the world we live in? Think about it. The world is constantly changing. Who will help make the changes of the future? Why not you? See what is wrong out there and try to fix it. Ask what you wish you had and figure out how to invent it.

5. Hang out with small children

Why does this matter? Because probably you are going to be a parent some day. Schools don't teach parenting skills. (I have no idea why not. Why do they think parenting isn't worth teaching?) Volunteer to help take care of kids in some way. You will learn a lot about them and about yourself.

6. Do some real science

Is there something you are curious about? Now is the time. Science isn't about memorizing facts they teach in school. Science is about investigation and discovery. Science is about finding evidence and causes. Do some real science. Investigate something. Think about the health of your parents, the habits of your dog, the growth of trees, water, airplanes, or cars. It doesn't matter what. Find out how they work. Figure out what might make them work better. This is real science.

7. Read

Yes. Read. Sounds like something your teacher would tell you. In this case they would be right. If the only thing you read all day is texts and web sites, you are not reading. Read something complicated. Read about something that requires logical arguments, gets your mind spinning, forces you to provide counter arguments, and makes you want to discuss it with people who know more than you do. It doesn't matter what you read about. It does matter that you think about something new in a careful reasoned way. And it also matters that you talk about what you have read.

8. Learn a language

Learning a new language can teach you a great deal about how your mind works and how your culture works. The only real way to learn one is to go somewhere where they speak that language and speak only that language for a while. When you are young it is easy to learn a language. Spend the summer learning a language and you will never regret it.

9. Meet someone new

I don't mean a new person who is like all the people you already know. Find someone from a different culture, a different world, who has nothing in common with you. Find out about them. Hang out with a different group. See the world from someone else's perspective. Everything you know about the world, everything you are most sure of, will be shaken up if you do this. This is a good thing.

10. Be bored

Sit quietly. Turn off all electronics. See what happens to your mind. Let it go while you are doing nothing – absolutely nothing – for an hour. You will be amazed at what happens when you shut it all down and let your mind wander. You will find out what you really think about things.

Try this stuff. It will put school in proper perspective for you.

Original Publication Dates

These articles originally appeared as blog posts on Education Outrage[1] and The Pulse: Education's Place for Debate (no longer online).

Chapter 1: Why?	
Why do we still have schools?	April 14, 2009
Why is school screwed up?	April 22, 2009
The 7th P	November 3, 2006
Why students cheat	July 13, 2010
Why do we teach what we teach?	February 25, 2009
Why students major in history and not science	June 9, 2009
Why don't we encourage schools to adapt to kids rather than the other way around?	November 3, 2014
Why do we give lectures? Why does anyone attend them?	January 24, 2015
Why not use the principles of prison reform to help schools?	August 18, 2013
Why our education system is the way it is: The Tillman Story	September 15, 2011
Why educate the elite? A lesson from Yale and George Bush while hanging out in Spain	September 18, 2013
The top ten mistakes in education: twenty years later	January 11, 2014

Chapter 2: Teaching the Wrong Things	
Wrong problem, wrong solution	December 15, 2006
In defense of what doesn't work	January 2, 2007
The curriculum deciders	October 22, 2010
Benito Juarez, Don Quixote, and Mexico City street kids	February 10, 2013
Stop pushing math in the name of reasoning	December 15, 2013
Why do we let the general public decide what should be taught in school?	September 30, 2013

1 http://educationoutrage.blogspot.com

Why textbooks suck	December 21, 2008
Sacred truths	August 30, 2010
An honest back to school message to students	September 3, 2013
NO to subjects and NO to requirements	June 20, 2011
The politics of history	March 20, 2009
The peril of passion	February 23, 2014
National standards are absurd	April 2, 2010
Does anyone care that kids all over the world think school is useless?	May 14, 2013

Chapter 3: Teaching the Wrong Way

Cavemen didn't have classrooms	September 26, 2007
The library metaphor	September 1, 2006
How to build a culture of illiteracy	November 13, 2010
The history teacher or the football coach?	September 18, 2006
Can you tell if it is school or prison?	February 25, 2010
Why can't we express emotion properly? A call for new school standards.	April 23, 2013
Teachers' despair: we cannot afford to be focused on training intellectuals	January 19, 2013
The myth of information retention	April 1, 2014
Exposure, cultural literacy and other myths of modern schooling	September 9, 2012
Reading and Math are very important, but I am not sure why: a message for those who can't see very well	August 6, 2015
Reading is no way to learn	July 11, 2015

Chapter 4: Technology Saves the World!

Games to the rescue!	April 27, 2013
Alex Trebek: hero of vocabulary preparation	December 6, 2009
The online learning disaster	June 16, 2014
Stanford decides to be Wal-Mart	June 2, 2014

Greed is not disruption	January 30, 2013
Princeton Professor teaches Coursera course; you must be kidding me!	November 28, 2011
Efficiency is not reform	October 1, 2014
MOOCs, the XPRIZE, and other things that will never change education	February 24, 2013
MOOCs: The New York Times gets it wrong again; Europe is not lagging behind the U.S.	November 19, 2012
Why are universities so afraid of online education?	November 20, 2013
More online nonsense: Starbucks and Arizona State agree to do nothing useful	May 27, 2009
Students: Be very afraid of online degree programs, especially if Pearson had anything to do with them	June 29, 2014
What online education should be	February 14, 2015

Chapter 5: Everyone Should Go To College?

Everyone must go to college (does anyone ever ask why?)	September 16, 2012
Should I go to a "hot" college?	December 17, 2009
The Big Lie: teaching never matters in university rankings	September 16, 2014
Stop cheating undergraduates of a useful education	April 11, 2014
How academic research has ruined our education system	February 9, 2010
Humanities are overrated	March 2, 2010
"What should I go to school for?"	August 12, 2009
Engaged learning	February 26, 2008
Confused about what college is about? So are colleges.	March 4, 2011
Preparing for a fictitious college	January 13, 2007
Only Harvard and Yale lawyers on the Supreme Court?	May 11, 2010
Please don't make me be a dentist!	December 12, 2011
Students: Life isn't a multiple choice test. Have some fun.	February 5, 2015
How do you know if you are college-ready?	May 24, 2015
Parents, relax. Your kid will get into college; the question is whether or not to go.	March 2, 2015

Chapter 6: Death to the Standardistas!

Death to the Standardistas!	February 11, 2008
What is wrong with trying to raise test scores?	July 19, 2009
Chinese do better on tests than Americans! Oh my God, what will we do?	December 13, 2010
The World Cup of testing	July 17, 2014
Frank Bruni thinks kids are too coddled. I think kids are too tested.	November 26, 2013
Measurement in preschool? Measure this!	August 25, 2013
"Life is a series of tests." What a load of nonsense.	September 13, 2010
American business hasn't a clue	November 1, 2007
Measure or die	March 7, 2013
Now I am disgusted!	March 18, 2007

Chapter 7: Dear Mr. Obama (and Other Politicians)

Mr. Obama wants big ideas? Here are 10 in education.	October 26, 2011
Duncan speaks; kids lose	November 19, 2009
A message to Bachmann, Duncan, and every other politician who thinks he or she knows how to fix education	August 18, 2011
Just the facts, ma'am	March 18, 2008
Thank you Arne, Bill, and Pearson for making this teacher so miserable	May 1, 2013
Why are you proud, Mr. Mayor?	June 15, 2007
More college graduates? Say it ain't so Mr. President.	March 3, 2009
Yes, Mr. Obama, money is the answer	September 22, 2010
What were your test scores, Mr. Obama?	March 10, 2009
The school of random facts	September 22, 2009
Hooray for the Democrats! Hooray for more accountability!	December 28, 2007
Why politicians and rich guys won't reform education	April 26, 2007
Public schools: where poor kids go to take tests	January 18, 2015

Duncan and Obama are actively preventing meaningful education change	June 15, 2009
Free community college? How about we fix high school, Mr. Obama?	January 9, 2015

Chapter 8: The New York Times (and others), Wrong Again

The New York Times and Nick Kristof want mass education. I want individualized education.	October 27, 2014
Spinning test prep into "choice"	August 30, 2009
The New York Times on the GED – wrong again	October 13, 2009
When the New York Times obsesses about math, every kid loses	July 27, 2014
Tom Friedman; wrong again, this time about education	November 21, 2010
Don't worry about Artificial Intelligence, Stephen Hawking	December 6, 2014
The misreporting of science by The New York Times and others	April 6, 2013
I translate Bill Gates dumb remarks on education	October 3, 2014
Bill Gates: wrong again about rating teachers	December 4, 2010
Madrassas, indoctrination, education, and Kristof	October 30, 2009
More absurdity from the New York Times and Nick Kristof	April 28, 2014
The fake choice: preschool or prison	October 30, 2013
Measuring teachers means education reform? You have got to be kidding!	May 7, 2011
Fared Zakaria and Ivy League graduates keep defending the liberal arts, but clearly the liberal arts didn't teach them to think	March 29, 2015

Chapter 9: Milo

My grandson visits	April 6, 2007
A drum should be a drum	October 15, 2007
Milo goes to kindergarten. We need to fix this fast.	September 22, 2010

Milo and the Park Slope parents meet Common Core; everybody loses (except Bill Gates and Big Pharma)	October 15, 2013
Milo's mom fights the Common Core	October 17, 2013
Milo takes a computer class and eventually will learn PowerPoint. Yippee!	June 1, 2013
Why can't school be more like camp?	July 26, 2015

Chapter 10: How To Fix Education

It's time for a change	November 12, 2014
A conversation about learning	November 18, 2014
We learn by talking	November 18, 2014
Persuasive arguments	November 23, 2014
Change just one thing	November 7, 2011
The old university system is dead – time for a professional university	January 31, 2014
What cognitive science tells us about how to organize school	October 17, 2009
Start by asking students what they want to learn	April 10, 2011
Pro-Choice: here's how to fix high schools	April 4, 2014
High school: Out with the old and in with the new	April 16, 2014
Cash in those chips	October 19, 2006
It's an old story – learning hasn't changed	November 29, 2012
How to fix the "STEM crisis"	March 6, 2011
To fix a country, fix education	June 14, 2011
A summer to-do list for students who hate (or love) high school	June 24, 2013

About the Author

Roger C. Schank, PhD

Dr. Schank founded the renowned Institute for the Learning Sciences at Northwestern University, where he is John P. Evans Professor Emeritus in Computer Science, Education and Psychology. He was Professor of Computer Science and Psychology at Yale University and Director of the Yale Artificial Intelligence Project. He was a visiting professor at the University of Paris VII, an Assistant Professor of Computer Science and Linguistics at Stanford University and research fellow at the Institute for Semantics and Cognition in Switzerland. He also served as the Distinguished Career Professor in the School of Computer Science at Carnegie Mellon University. He is a fellow of the AAAI and was founder of the Cognitive Science Society and co-founder of the Journal of Cognitive Science. He holds a PhD in Linguistics from University of Texas.

In 1994, he founded Cognitive Arts Corporation, a company that designs and builds high quality multimedia simulations for use in corporate training and for online university-level courses. The latter were built in partnership with Columbia University.

In 2002 he founded Socratic Arts, a company that is devoted to making high quality e-learning affordable for both businesses and schools.

He is the author of more than 20 books on learning, language, artificial intelligence, education, memory, reading, e-learning, and story telling. The most recent are Make School Meaningful – and Fun!, Virtual Learning, Coloring Outside the Lines: Raising a Smarter Kid by Breaking All the Rules, Scrooge meets Dick and Jane, Engines for Education, and Designing World Class E-Learning.

Roger Schank's website is at: www.rogerschank.com

The Education Outrage blog can be found at: educationoutrage.blogspot.com

Also from Constructing Modern Knowledge Press

Invent To Learn: Making, Tinkering, and Engineering in the Classroom

by Sylvia Libow Martinez and Gary Stager

Join the maker movement! There's a technological and creative revolution underway. Amazing new tools, materials and skills turn us all into makers. Using technology to make, repair or customize the things we need brings engineering, design and computer science to the masses. Fortunately for educators, this maker movement overlaps with the natural inclinations of children and the power of learning by doing. The active learner is at the center of the learning process, amplifying the best traditions of progressive education. This book helps educators bring the exciting opportunities of the maker movement to every classroom.

The Invent To Learn Guide to 3D Printing in the Classroom: Recipes for Success

by David Thornburg, Norma Thornburg, and Sara Armstrong

This book is an essential guide for educators interested in bringing the amazing world of 3D printing to their classrooms. Learn about the technology, exciting powerful new design software, and even advice for purchasing your first 3D printer. The real power of the book comes from a variety of teacher-tested step-by-step classroom projects. Eighteen fun and challenging projects explore science, technology, engineering, and mathematics, along with forays into the visual arts and design. The Invent To Learn Guide to 3D Printing in the Classroom is written in an engaging style by authors with decades of educational technology experience.

Visit CMKPress.com for more information

Sylvia's Super-Awesome Project Book: Super-Simple Arduino (Volume 2)

by Sylvia (Super-Awesome) Todd

In this super fun book, Sylvia teaches you to understand Arduino microcontroller programming by inventing an adjustable strobe and two digital musical instruments you can play! Along the way, you'll learn a lot about electronics, coding, science, and engineering.

Written and illustrated by a kid, for kids of all ages, Sylvia's whimsical graphics and clever explanations make powerful STEM (Science, Technology, Engineering, and Math) concepts accessible and fun.

The Invent To Learn Guide to Fun

by Josh Burker

The Invent To Learn Guide to Fun features an assortment of insanely clever classroom-tested "maker" projects for learners of all ages. Josh Burker kicks classroom learning-by-making up a notch with step-by-step instructions, full-color photos, open-ended challenges, and sample code. Learn to paint with light, make your own "Operation Game," sew interactive stuffed creatures, build "Rube Goldberg" machines, design artbots, produce mathematically generated mosaic tiles, program adventure games, and more! Your MaKey MaKey, LEGO, old computer, recycled junk, and 3D printer will be put to good use in these fun and educational projects. With The Invent To Learn Guide to Fun in hand, kids, parents, and teachers are invited to embark on an exciting and fun learning adventure!